LOGIC IN WONDERLAND

An Introduction to Logic through Reading
Alice's Adventures in Wonderland

Teacher's Guidebook

LOGIC IN WONDERLAND

An Introduction to Logic through Reading
Alice's Adventures in Wonderland

Teacher's Guidebook

Nitsa Movshovitz–Hadar
Technion — Israel Institute of Technology, Israel

Atara Shriki
Oranim Academic College of Education, Israel

WS Education

NEW JERSEY · LONDON · SINGAPORE · BEIJING · SHANGHAI · HONG KONG · TAIPEI · CHENNAI · TOKYO

Published by

WS Education, an imprint of
World Scientific Publishing Co. Pte. Ltd.
5 Toh Tuck Link, Singapore 596224
USA office: 27 Warren Street, Suite 401-402, Hackensack, NJ 07601
UK office: 57 Shelton Street, Covent Garden, London WC2H 9HE

British Library Cataloguing-in-Publication Data
A catalogue record for this book is available from the British Library.

LOGIC IN WONDERLAND
An Introduction to Logic through Reading *Alice's Adventures in Wonderland* **— Teacher's Guidebook**

ISBN 978-981-3208-62-9
ISBN 978-981-3209-81-7 (pbk)

For any available supplementary material, please visit
https://www.worldscientific.com/worldscibooks/10.1142/10409#t=suppl

Desk Editors: V. Vishnu Mohan/Kwong Lai Fun

Typeset by Stallion Press
Email: enquiries@stallionpress.com

Printed in Singapore by Mainland Press Pte Ltd.

Nitsa: To my grandsons: Amit (9), Michael (12), Gur (12);
grand-teenagers: Yonatan (15), Saar (16), Gideon (17);
grand-"ladies": Shoval (19), Afik (20), Maya (21).

Atara: To my husband: Meir;
my children: Guy, Roy, Noa, Keren;
and my granddaughter: May.

Preface

Ever since the appearance of Lewis Carroll's *Alice's Adventures in Wonderland* more than 150 years ago, this story has remained one of the most popular children's books of all time. From the time of its publication in 1865 and throughout the years since then, it has been the subject of numerous interpretations from a wide variety of perspectives, including philosophical, religious, psychological and scientific viewpoints. The book has been translated into more than 150 languages, and has served as inspiration for screenwriters, playwrights, musicians, authors, and artists from various disciplines. Multiple websites are dedicated to the book and its author, with online discussion groups actively engaging in dialogue. People of all ages are touched by this exceptional story, finding in it many elements relevant to their own lives.

In our own professional lives we have found that logic is one of the least favorite subjects in mathematics among pre-service teachers as well as among in-service ones. They view it as a dry and tedious subject, devoid of practical use, one that is virtually impossible to incorporate into the school curriculum. It has been well established, however, that logic plays an important role in the foundation of mathematics. Hence, we found ourselves faced with a challenge — how can we transform the instruction of logic into a stimulating and appealing area of study? How can we arouse awareness in students to the pertinence of logic in the school curriculum? These questions, together with the knowledge that Charles Lutwidge Dodgson (better known by his pen name Lewis Carroll), the author of *Alice's Adventures in Wonderland,* taught logic at Oxford University, led us to consider the possibility of developing a logic course based on his book. After a thorough analysis of the text, we found that we indeed could use *Alice* to develop a logic course, the essence of which emerges naturally from the sequences of fantasy embodied in its plot. We were further inspired by Martin Gardner's

The Annotated Alice,[1] which primarily contains notes and interpretations addressing social and personal features of Lewis Carroll himself. Nevertheless, we found in it elements worthy of reference, although we did not find in it reference to logic *per se*.

This book is based on our experience acquired from the logic courses we taught at Oranim Academic College of Education in Israel as part of training elementary school mathematics teachers. This practical experience was accompanied by research we carried out at Technion — Israel Institute of Technology. The first version of this book was published in 2012 in Hebrew by Mofet Institute — a Center for the Research, Curriculum and Program Development in Teacher Education in Israel. It was the feedback received from the local professional community that encouraged us to translate the book into English.

We believe that not only mathematics teachers, but Lewis Carroll's admirers as well as math enthusiasts of various kinds, will find this book challenging and stimulating.

About the Learning Environment

This book is intended to serve as a basis for the learning environment in an introductory course in mathematical logic, based on reading the entire first chapter of *Alice's Adventures in Wonderland* as well as of additional excerpts from other chapters of that book. Although this course touches upon many topics of propositional calculus and predicate logic, it should not be viewed as a substitute for a comprehensive course in mathematical logic.

Design of the learning environment was carried out based on the four steps described in detail in the paper written by Movshovitz-Hadar & Shriki (2009).[2]

Each chapter of this book, except for the last one which is an overall summary, is dedicated to a single topic. It includes the following: (i) The logical-mathematical background related to the particular topic, intended for presentation by the instructor; (ii) A series of students' worksheets tailored to the topic. These

---

[1] Lewis Carroll, *The Annotated Alice,* The definitive edition with introduction and notes by Martin Gardner. W.W. Norton & Company, Inc., NY, 2000.

[2] Additional details on designing a learning environment appear in: Movshovitz-Hadar, N. & Shriki, A. (2009). Logic in Wonderland — *Alice's Adventures in Wonderland* as a context of a course in logic for future elementary teachers. In: B. Clarke, R. Millman & B. Grevholm (Eds.), *Tasks in Primary Mathematics Teacher Education, Use and Exemplars* (pp. 85–103). Springer Mathematics Teacher Education Series, Vol. 4, Andrea Peter-Koop (Ed.), Springer U.S. http://www.springerlink.com/content/qxp0161160086m72/?p=df6abc02380d412f94c99c8aad3afab2&pi=5.

worksheets are designed for students' independent work or group work and include suggested answers to each question or task in *Handwriting font*. Note: The worksheets, without solutions, appear in a separate Student's Workbook. (iii) Detailed suggestions for navigating the class through the learning process. In particular the first worksheet in each chapter is an opening exercise that also repeats itself as the summary worksheet. The worksheet itself and the expected answers to the questions that appear in it are presented only once at the end of each chapter in the Teacher's Guidebook. In the Students' Workbook, this work-sheet appears twice — once as the opening exercise and another time as a summary exercise.

In addition, the Students' Workbook includes in each chapter a final worksheet containing instructions for student's reflection upon the change in his or her knowledge as a result of the study of the specific topic of that chapter. This is done through comparing his or her answers to the opening and the summary exercises. This final worksheet does not appear in the Teacher's Guidebook.

It is recommended to open the instruction of each chapter by asking the students to relate to the opening exercise. Upon its completion it is advisable to explain to the students that they will return to this page at the end of the chapter, and then they will be able to compare their initial knowledge with the knowledge they acquired during the course of learning that chapter. This will enable reflection on the change in their knowledge and on points that still require clarification. Therefore, it is recommended not to discuss the answers to the opening exercise immediately after its completion, but to carry on the work as indicated in each chapter.

The last chapter in this book includes summary tables and ten worksheets that relate to all the topics dealt with in Chapters 1–9 of the book.

We hope that student teachers and other learners will enjoy this unique learning environment and develop a positive attitude to mathematical logic. Readers and users of the book are encouraged to send us by e-mail comments about their experiences.

Nitsa
nitsa@technion.ac.il

Atara
atarashriki@gmail.com

July 2018

Acknowledgments

We wish to thank the publisher of the original Hebrew version, Mofet Institute — a Center for the Research, Curriculum and Program Development in Teacher Education, for providing the permission for publishing an English version of the original Hebrew version of the book.

Thanks are also due to Academic Language Experts and to Debbie and Shlomo Libeskind for their assistance in the translation and adaptation of the Hebrew edition to the English one.

All excerpts from the book and references to them are taken from *Alice's Adventures in Wonderland* in e-book form at Project Gutenberg as appeared in the "no rights reserved" website: http://www.gutenberg.org/files/28885/28885-h/ 28885-h.htm (Retrieved August 2017). All the illustrations by John Tenniel are taken from the "no rights reserved" website: https://commons.wikimedia.org/ wiki/Category:John_Tenniel%27s_illustrations_of_Alice%27s_Adventures_in_ Wonderland. Other illustrations are from: http://www.freepik.com.

Contents

Chapter 1

The Logical Connective AND

Alice AND the Rabbit are
the Heroes of the Story

1.1. A Look Inside the Text

The lesson opens by reading the first paragraph of the book *Alice's Adventures in Wonderland*, on page 1, which appears in boldface below.

Down the Rabbit-Hole

 LICE was beginning to get very tired of sitting by her sister on the bank, and of having nothing to do: once or twice she had peeped into the book her sister was reading, but it had no pictures or conversations in it, "and what is the use of a book," thought Alice, "without pictures or conversations?"

So she was considering in her own mind (as well as she could, for the hot day made her feel very sleepy and stupid) whether the pleasure of making a daisy-chain would be worth the trouble of getting up and picking the daisies, when suddenly a White Rabbit with pink eyes ran close by her.

There was nothing so *very* remarkable in that; nor did Alice think it so *very* much out of the way to hear the Rabbit say to itself, "Oh dear! Oh dear! I shall be too late!" (when she thought it over afterwards, it occurred to her that she ought to have wondered at this, but at the time it all seemed quite natural); but when the Rabbit actually *took a watch out of its waistcoat-pocket*, and looked at it, and then hurried on, Alice started to her feet, for it flashed across her mind that she had never before seen a rabbit with either a waistcoat-pocket, or a watch to take out of it, and burning with curiosity, she ran across the field after it, and was just in time to see it pop down a large rabbit-hole under the hedge.

1.2. Pedagogic Discussion — The Importance of Reading

Before beginning the lesson, the passage above may be used to initiate a student discussion on the importance of reading and of encouraging children to read.

Below are sample questions that can be used in the discussion:

- What books do children like to read? Why?

- Children's reading habits have changed in recent years with increasing availability of visual materials. What measures may be taken to help them cultivate a love of reading?

Take the opportunity to browse through *Alice's Adventures in Wonderland* and notice that it is full of illustrations.

Remark: We will return to this passage in Chapter 2, where we will address the connective NOT.

1.3. Worksheet 1a: Opening Exercise. Lead-in to the Logical Connective AND

Instruction of this chapter begins with Worksheet 1a: Opening Exercise (hereforth abb. WS1a) to be completed as individual practice by each student. Upon completion of the task, notify the students that they will return to this worksheet at the end of the chapter (Worksheet 1d Summary Exercise); then they will be able to compare their initial understanding with that which they acquired during the course of the chapter, while reflecting upon the changes in their understanding along with the points where further clarification is still required (Worksheet 1e, which appears only in the Student's Workbook). For this reason it is recommended that discussion of this worksheet will be held off until the end of the chapter. In some cases, the teacher may prefer to collect WS1a and give it back only upon completion of WS1d when students are ready to work on WS1e.

Because this lead-in worksheet is repeated again as a summary exercise, the sheet (together with proposed solutions) appears only once — at the end of this chapter.

1.4   A Look Inside the Text

Read the second paragraph of *Alice's Adventures in Wonderland* on page 1, which appears in boldface below. This paragraph will serve as the basis for discussion of the logical connective AND.

Down the Rabbit-Hole

 LICE was beginning to get very tired of sitting by her sister on the bank, and of having nothing to do: once or twice she had peeped into the book her sister was reading, but it had no pictures or conversations in it, "and what is the use of a book," thought Alice, "without pictures or conversations?"

So she was considering in her own mind (as well as she could, for the hot day made her feel very sleepy and stupid) whether the pleasure of making a daisy-chain would be worth the trouble of getting up and picking the daisies, when suddenly a White Rabbit with pink eyes ran close by her.

There was nothing so *very* remarkable in that; nor did Alice think it so *very* much out of the way to hear the Rabbit say to itself, "Oh dear! Oh dear! I shall be too late!" (when she thought it over afterwards, it occurred to her that she ought to have wondered at this, but at the time it all seemed quite natural); but when the Rabbit actually *took a watch out of its waistcoat-pocket*, and looked at it, and then hurried on, Alice started to her feet, for it flashed across her mind that she had never before seen a rabbit with either a waistcoat-pocket, or a watch to take out of it, and burning with curiosity, she ran across the field after it, and was just in time to see it pop down a large rabbit-hole under the hedge.

1.5. Worksheet 1b: The Logical Connective AND

The central topic of this chapter — the logical connective AND — kicks off with Worksheet 1b, which begins with the excerpt shown above (Section 1.4).

Remarks:

- Worksheet 1b is to be completed by students in pairs or in small groups to allow for consultation.
- It is suggested that upon completion of the worksheet, students present their answers to the class, in order to initiate a whole class discussion of the solutions.
- At this point it is advised to avoid any judgment of students' answers; points of disagreement and misperceptions, however, should be noted.
- After presentation of the logical and mathematical background by the teacher (Section 1.6 below), students' answers should be revisited and discussed once again.

Worksheet 1b and Proposed Solutions

1. What are the two things that happened to Alice as a consequence of it being a hot day?

    The hot day made Alice:
    1. Very sleepy   2. Stupid.

2. What are the two activities that Alice must trouble herself with in order to make a daisy-chain?

    In order for Alice to make a daisy-chain, she must:
    1. Get up   2. Pick the daisies.

3. What syntactic component helped you answer the preceding two questions?

    The word "and" in the phrases "sleepy and stupid," and "getting up and picking the daisies."

4. Could the sentence: "When suddenly a White Rabbit with pink eyes ran close by her" be phrased instead as "when suddenly a rabbit that was white and had pink eyes ran close by her," without changing its meaning? If not, why not? If so — do you think that there is no difference between "with" and "and"?

    In spoken language there is no significant difference between the two forms. In logic, however, the term "with" is not considered a valid substitute for the term "and."

1.6. Logical and Mathematical Background

Upon completion of Worksheet 1b and after discussion of the students' answers, the following concepts and topics are to be presented:

A.   A simple statement and the concept of truth value
B.   The Use of the logical connective AND to obtain a compound statement
C.   A truth table
D.   The Use of the logical connective AND in daily language

To ease the learning process some of the concepts will be introduced using examples.

A.   A simple statement and the concept of truth value

A **statement** is a sentence in spoken or written language that makes a declaration (in contrast to an interrogative or exclamatory sentence), and commonly ends with a period (in contrast to a question mark or an exclamation mark).

A statement is differentiated by the ability to pose the following question with regard to it: "Is its content TRUE?" In the event that the answer is known, it will either be "its content is TRUE" or "its content is FALSE." A sentence about which this question cannot be posed is not a statement. For example: "Did it rain yesterday?" is not a statement, as the question: "Is it true that did it rain yesterday?" cannot be asked. Yet "It rained yesterday" is a statement, as the truth of its content can be verified, and the question "Is it true that it rained yesterday?" can be answered in the affirmative or in the negative. Similarly, the sentences "Rain, rain, go away" and "Let us pray for rain" are not statements, because it is meaningless to ask about the truth of their content. However, "tomorrow it will rain" is a statement because the truth of its content can be examined. Even if not available at a particular moment in time, the answer, when it does become available, will be either TRUE or FALSE. In mathematics, statements are always stated in present tense.

Additional examples: "$3 + 2 = 5$" is a statement, while "Is seven a prime number?" is not a statement, because it is interrogative and not declarative.

The **truth value** of a statement, therefore, is **TRUE** or **FALSE**. By convention the truth value TRUE is denoted by the letter T and the truth value FALSE is denoted by the letter F. Each statement has one and only one truth value.[1]

Keep in mind, however, that the truth value of a statement depends at times on the context in which it is stated. For example, to determine the truth value of "Today is Sunday" it is crucial to know on what day of the week the statement was made.

A **simple statement**, like any statement, is a declarative sentence consisting of a subject and a predicate; it may also contain a description or adjective, but not more than that. A simple statement may not contain words such as NOT, OR, and AND — words that are conventionally called "connectives" (more details on NOT and OR appear in Chapters 2 and 3 respectively). It also may not include words like ALL and THERE EXISTS — words that are conventionally known as "quantifiers" (more details appear in Chapter 7). Thus, for example, "Today it is snowing in the mountains" and "Michael is going to the beach" are simple statements. But "Today it is snowing in the mountains or Michael is going to the beach" is not a simple statement. Also, "Michael is not going to the beach" is not a simple statement. It is the negation of a simple statement. Simple statements are the building blocks of logic, its "atoms." By convention, simple statements are designated by lower case letters such as p and q, and each is addressed as an independent unit. For example, p could represent the statement: "In the triangle before you the measure of the interior angle α is 35°."

B. The use of the logical connective AND to obtain a compound statement

A compound statement may be formed from two simple statements or their negation, through the use of the logical connectives AND and OR. In this lesson, we will focus on the logical connective AND. In Chapter 3 of this book, we will address the logical connective OR.

---

[1] The term "one and only one" is not redundant, but rather is a precise way to express the fact that there exists exactly one truth value. "One" implies at least one, i.e. not less than one, while "only one" implies at most one, i.e. not greater than one.

The following is an example of use of the logical connective AND: "It is raining today in the mountains and Michael has flown to Europe."

The logical connective AND is a binary logical operator. The term "logical operator" indicates that its operand is a statement, while "binary" indicates that it operates on two statements, deriving one new statement from them. When two simple statements are linked using the binary logical connective AND, the result is a compound statement, the truth value of which depends upon the truth values of its two component statements. Its truth value is TRUE if the truth value of each of its component statements is TRUE. If the truth value of at least one of its component statements is FALSE (that is, one of the component statements or both of them have the truth value FALSE), then the compound statement has truth value FALSE. Determining the truth value of a compound statement by the truth value of the simple statements from which it is formed is called **propositional calculus**. The conventional notation for the statement p AND q is $p \land q$.

C. A truth table

A **truth table** is a table used to present all possible truth values that a specific statement (simple or compound) could assume. In the case of a compound statement, the truth value is determined in accordance with the possible values of its component statements. In effect, truth tables show the results of applying "operations" to statements. This lesson deals with the "operation" AND. In a truth table, statements are denoted by lower case letters, and truth values by T (for a TRUE statement) and F (for a FALSE statement). Below is a truth table for two statements, p and q, and the compound statement derived from them by using the logical connective AND: $p \land q$.

Truth table for the logical connective AND

p	q	$p \land q$
T	T	T
T	F	F
F	T	F
F	F	F

Note: If a compound statement is formed from two simple statements using the logical connective AND, then the number of possible combinations of truth values of the component simple statements is $2^2 = 4$. Similarly, if it is formed from n simple statements, then the number of possible combinations is 2^n. For example:

p	q	r	$p \wedge q \wedge r$
T	T	T	T
T	T	F	F
T	F	T	F
T	F	F	F
F	T	T	F
F	T	F	F
F	F	T	F
F	F	F	F

D. The use of the logical connective AND in daily language

1. Use of the logical connective AND in spoken language is often abbreviated through omission of the subject of the second sentence. Thus, for example, the statement "This boy is tall and thin" is commonly understood as a shortened form of the statement "This boy is tall AND this boy is thin", which in turn is composed of the two simple statements "This boy is tall" and "This boy is thin."

2. Additional words that suggest the meaning AND exist in the spoken language. For example:

   • The statement "Not much rain fell yet there is a lot of mud" is identical in meaning to the statement "Not much rain fell and there is a lot of mud."

   • The statement "My car is beautiful but small" is identical in meaning to the statement "My car is beautiful and small."

Logically, words like "but", "yet" and "still" are generally equivalent in meaning to the word "and." Nevertheless, in spoken language they serve to emphasize or to contrast.

3. In logic, the meaning of the compound statement $p \wedge q$ is identical to the meaning of the compound statement $q \wedge p$. In spoken language, however, the word "and" is sometimes used to indicate cause and effect. For example, although the statement "The dog barked and the girl cried" is logically equivalent to the statement "The girl cried and the dog barked," in spoken language we usually attribute the girl's crying to the barking of the dog in the former, while in the latter we attribute the dog's barking to the girl crying.[2]

1.7. Review of Answers to Worksheet 1b

After students have become familiar with the logical background, it is advised to ask them to go back and review their answers to Worksheet 1b and correct them if necessary.

1.8. Worksheet 1c: The Logical Connective AND — Inference Using a Truth Table

Worksheet 1c addresses inferences from statements containing the logical connective AND by using truth tables.

Remarks:

• This worksheet may be completed as individual practice, or in pairs or small groups with group discussion encouraged.
• Upon completion of the worksheet, it is recommended that the students present their answers to the class to initiate a whole class discussion of the solutions.
• In answering the questions to this worksheet the following rule should be kept in mind: If two simple statements are linked using the binary logical connective AND, the result is a compound statement the truth value of which depends on the truth values of its two components. The truth value of the compound statement is thus indicated in the truth table for the logical connective AND.

---

[2] Logical equivalence is addressed in more details in Chapter 4.

Worksheet 1c and Proposed Solutions

The White Rabbit that Alice saw in Wonderland was confused and somewhat nervous. It was nervous primarily because it was afraid of the Queen of Hearts who ruled Wonderland. When the Rabbit first met Alice, it dropped the white kid gloves and the fan that it was holding. After a time Alice heard a little pattering of footsteps in the distance and looked up eagerly to see who was coming (page 35).

Alice Meets the White Rabbit Once Again

 It was the White Rabbit, trotting slowly back again, and looking anxiously about as it went, as if it had lost something; and she heard it muttering to itself, "The Duchess! The Duchess! Oh my dear paws! Oh my fur and whiskers! She'll get me executed, as sure as ferrets are ferrets! Where can I have dropped them, I wonder?"
Alice guessed in a moment that it was looking for the fan and the pair of white kid gloves, and she very good-naturedly began hunting about for them, but they were nowhere to be seen.

Each item below contains two statements followed by a question.
Circle the correct answer and explain your reasoning based on the given information.

1. Both of the following statements are TRUE in Wonderland:

    (i) The White Rabbit trotted slowly back again, and it looked anxiously about as it went.
    (ii) The White Rabbit looked anxiously about as it went.

    Which of the following is correct?

    (A.) **The statement "The White Rabbit trotted slowly back again" is TRUE.**

    B. The statement "The White Rabbit trotted slowly back again" is FALSE.

    C. The two given statements contradict one another.

    D. There is insufficient information given to determine whether the White Rabbit trotted slowly back again.

    Reasoning:

    Let us designate the statements as follows:

    p: *The White Rabbit trotted slowly back again.*

    q: *The White Rabbit looked anxiously about as it went.*

p	q	$p \wedge q$
T	T	T
T	F	F
F	T	F
F	F	F

    Given: The compound statement $p \wedge q$ is TRUE. In accordance with the truth table for the logical connective AND, the first row matches the given information; thus we obtain that statement p must be TRUE, and that statement q must be TRUE as well.

    It is also given that q is TRUE. Our conclusion is that statement p is necessarily TRUE. This corresponds to option A, which has been marked as the correct answer.

    This question may be answered without using a truth table. Since the compound statement $p \wedge q$ is TRUE, both of its components — the two simple statements — must be TRUE as well. q is given as TRUE, p must be TRUE as well.

2. The following statement is FALSE in Wonderland:
    "The White Rabbit is looking for its white kid gloves and the White Rabbit is looking for its shoes."

    The following statement is TRUE in Wonderland:
    "The White Rabbit is looking for its white kid gloves."

    Which of the following is correct?

    A. The statement "The White Rabbit is looking for its shoes" is TRUE.

    B. **The statement "The White Rabbit is looking for its shoes" is FALSE.**

    C. The two given statements contradict one another.

    D. There is insufficient information given to determine whether the White Rabbit is looking for its shoes.

    Reasoning:

    Let us designate the statements as follows:

    p: *The White Rabbit is looking for its white kid gloves.*

    q: *The White Rabbit is looking for its shoes.*

    Let us construct a truth table for the logical connective AND.
    Given: The compound statement $p \wedge q$ *is FALSE.*

p	q	$p \wedge q$
T	T	T
T	F	F
F	T	F
F	F	F

    In accordance with the truth table for the logical connective AND, rows 2–4 match the given information.

    From these three rows we obtain that each one of the statements p and q could be either TRUE or FALSE.

    It is also given that p is a TRUE statement.

    Hence, from the three rows we marked previously, we are left with only the second row of the table.

p	q	$p \wedge q$
T	T	T
T	F	F
F	T	F
F	F	F

Therefore we obtain that the statement q must be FALSE. This corresponds to option B, which has been marked as the correct answer.

The question may also be answered without using a truth table. Since the compound statement p ∧ q is FALSE, at least one of its two components — the simple statements p or q — is FALSE. Because p is TRUE, it follows that q must be FALSE.

3. Both of the following statements are FALSE in Wonderland:

    (i) Alice is looking for the Duchess and the White Rabbit is looking for the white kid gloves.
    (ii) Alice is looking for the Duchess.

    Which of the following is correct?

    A. The statement "The White Rabbit is looking for the white kid gloves" is TRUE.

    B. The statement "The White Rabbit is looking for the white kid gloves" is FALSE.

    C. The two given statements contradict one another.

    D. **There is insufficient information given to determine whether the White Rabbit is looking for the white kid gloves.**

Reasoning:

Let us designate the statements as follows:

p: Alice is looking for the Duchess.

q: The White Rabbit is looking for the white kid gloves.

Given: The compound statement p ∧ q is FALSE.

In accordance with the truth table for the logical connective AND, rows 2-4 match the given information.

p	q	$p \wedge q$
T	T	T
T	F	F
F	T	F
F	F	F

From these three rows we obtain that each one of the statements p and q could be either TRUE or FALSE.

It is also given that p is FALSE.

Thus, we are left with rows 3 and 4 of the table.

p	q	$p \wedge q$
T	T	T
T	F	F
F	T	F
F	F	F

Hence we conclude that the statement q can be either TRUE or FALSE;

That is, we cannot determine whether the White Rabbit is looking for the white kid gloves. This corresponds to option D, which has been marked as the correct answer.

The question may also be answered without using a truth table. Since the compound statement p ∧ q is FALSE, at least one of its two components — the simple statement p or the simple statement q — is FALSE. Since p is FALSE, we cannot determine whether q is TRUE, or whether it is FALSE as well.

4. In Wonderland the following statement is TRUE:
    "The White Rabbit is looking for the fan and the White Rabbit likes the Duchess."

    In Wonderland the following statement is FALSE:
    "The White Rabbit likes the Duchess."

    Which of the following is correct?

    A.   The statement "The White Rabbit is looking for the fan" is TRUE.

    B.   The statement "The White Rabbit is looking for the fan" is FALSE.

    C.   **The two given statements contradict one another.**

    D.   There is insufficient information given to determine whether the White Rabbit is looking for the fan.

Reasoning:

Let us designate the statements as follows:

p: The White Rabbit is looking for the fan.

q: The White Rabbit likes the Duchess.

Given: The compound statement $p \wedge q$ is TRUE.

In accordance with the truth table of the logical connective AND, the first row matches the given information.

p	q	$p \wedge q$
T	T	T
T	F	F
F	T	F
F	F	F

From this row we obtain that statement p must be TRUE and that statement q must be TRUE as well.

It is given, however, that q is FALSE. Therefore we conclude that the two given statements contradict one another. This corresponds to option C, which has been marked as the correct answer.

The question may also be answered without using a truth table. Since the compound statement $p \wedge q$ is TRUE, our conclusion is that its two components — the simple statements p and q — are necessarily TRUE. Since q is FALSE, we conclude that the given information is contradictory.

5. In Wonderland the following statement is FALSE:
    "The White Rabbit is looking for the fan and the White Rabbit likes the Duchess."

    In Wonderland the following statement is TRUE:
    "The White Rabbit is looking for the white kid gloves."

    Which of the following is correct?

    A. The statement "The White Rabbit is looking for the fan and the white kid gloves" is TRUE.

    B. The statement "The White Rabbit is looking for the fan and the white kid gloves" is FALSE.

    C. The two statements contradict one another.

    D. **There is insufficient information given to determine whether the White Rabbit is looking for the fan and the white kid gloves.**

Reasoning:

Let us designate the statements as follows:

p: The White Rabbit is looking for the fan.

q: The White Rabbit likes the Duchess.

r: The White Rabbit is looking for the white kid gloves.

Given: The compound statement p ∧ q is FALSE. In accordance with the truth table for the logical connective AND, rows 3-8 match the given information.

p	*q*	*r*	*p ∧ q*	*p ∧ r*
T	T	T	T	T
T	T	F	T	F
T	F	T	F	T
T	F	F	F	F
F	T	T	F	F
F	T	F	F	F
F	F	T	F	F
F	F	F	F	F

It is also given that r is TRUE. Thus, we are left with rows 3, 5 and 7 of the table.

From these rows we obtain that p ∧ r can be either TRUE or FALSE. We conclude that it cannot be determined whether the White Rabbit is looking for the fan and the white kid gloves. This corresponds to option D, which has been marked as the correct answer.

p	*q*	*r*	*p ∧ q*	*p ∧ r*
T	T	T	T	T
T	T	F	T	F
T	F	T	F	T
T	F	F	F	F
F	T	T	F	F
F	T	F	F	F
F	F	T	F	F
F	F	F	F	F

1.9. Worksheets 1d and 1e: Summary Exercise for the Logical Connective AND

The summary exercise is composed of two Worksheets: 1d and 1e.

Remarks:

- Worksheet 1d is to be performed as individual practice by the students. In Worksheet 1e (which appears in the students' workbook only) students will compare their current answers with the answers they gave at the start of the chapter (Worksheet 1a).
- Upon completion of the summary exercise, it is recommended that a whole class discussion be held in order to consider the changes that have taken place in students' understanding and perceptions through the course of this chapter, as well as to identify the particular difficulties encountered with the subject matter.

Worksheet 1d and Proposed Solutions

Answer the questions in this worksheet in their entirety, providing as much detail as possible. If necessary, you may indicate: "I did not understand the question, therefore I have not answered it." Make sure not to go back to Worksheet 2a before you complete your work on this worksheet.

1.   What are the differences between the following three sentences:

   A.   John went to school today.

   B.   Did John go to school today?

   C.   Go to school, John!

   The first sentence is declarative (it describes a fact), the second is interrogative and the third is exclamatory (an instruction, in this case). Only with regard to the first of the three can the question "Is the content TRUE or FALSE?" be posed. It cannot be asked of the interrogative, nor can it of the exclamatory.

2.   Provide three mathematical statements where the differences between them are similar to the differences between the sentences in Question 1.

   A.   *The sum of the angles in a triangle in the Euclidean plane is 180 degrees (TRUE); a second-order equation with real coefficients has three solutions (FALSE).*

   B.   *Is every prime number greater than 10?*

   C.   *Divide 27 by 3!*

3.   What are the differences between the following three sentences:

   A.   There are seven days in the week.

   B.   The month which follows January is August.

   C.   There exists a star in the universe other than EARTH where two-legged beings live.

Each of these three sentences describes a fact, but the first is TRUE, the second is FALSE and the third is indeterminate (at present), that is, we do not know whether it is TRUE or FALSE.

4.   Provide three mathematical statements where the differences between them are similar to the differences between the sentences in Question 3.

   A.   *The number of diagonals from one vertex of a hexagon is 4.*

   B.   *Four is a prime number.*

   C.   *Every even number greater than 4 may be represented as the sum of two prime odd numbers. This is Goldbach's Conjecture, which, as of this writing, has been neither proven nor refuted.*

5. The following instruction was given in a classroom: "All those who attend the extracurricular activities: gymnastics and computers, please raise their hand." Let us consider the following four students:

   - Jane attends gymnastics and computers.

   - Jonathan does not attend gymnastics and does not attend computers.

   - Susan doesn't attend gymnastics but does attend computers.

   - Adam attends gymnastics but does not attend computers.

   Who of the four should raise their hand?

   A. All four (why?)

   B. Three of them (who? why?)

   C. Two of them (who? why?)

   D. Only one (who? why?)

   E. None of them (why?)

   The wrong answer, but the one which is most likely, is b. Many would expect that the question is aimed at those who participate in either of the two extracurricular activities, not necessarily both of them. Yet it is reasonable to assume that even those who think so, think that those who participate in both should also raise their hand. The correct answer is D. Only those who participate in both extracurricular activities should raise their hand, since the question was worded such that the AND connects between the two activities. Based on the given information, Jane is the one who attends both activities. If the question was intended to determine those who attend at least one of the extracurricular activities, the sentence should have been worded accordingly (for more details on the logical Connective OR see Chapter 3).

Chapter 2

The Logical Connective NOT

Alice is NOT the Rabbit

2.1 Worksheet 2a: Opening Exercise. Lead-in to the Logical Connective NOT

Instruction of this chapter begins with Worksheet 2a: Opening Exercise (hereforth abb. WS2a) to be completed as individual practice by each student. Upon completion of the task, notify the students that they will return to this worksheet at the end of the chapter (Worksheet 2e Summary Exercise); then they will be able to compare their initial understanding with that which they acquired during the course of the chapter, while reflecting upon the changes in their understanding along with the points where further clarification is still required (Worksheet 2f, which appears only in the Student's Workbook). For this reason it is recommended that discussion of this worksheet will be held off until the end of the chapter. In some cases the teacher may prefer to collect WS2a and give it back only upon completion of WS2e when students are ready to work on WS2f.

Because this lead-in worksheet is repeated again as a summary exercise, the sheet (together with proposed solutions) appears only once — at the end of this chapter.

2.2. A Look Inside the Text

Read the first lines of the third paragraph on page 1 of the book *Alice's Adventures in Wonderland*, which appear in boldface print below. These lines will serve as the basis for discussion of the concept 'exception.'

Down the Rabbit-Hole

LICE was beginning to get very tired of sitting by her sister on the bank, and of having nothing to do: once or twice she had peeped into the book her sister was reading, but it had no pictures or conversations in it, "and what is the use of a book," thought Alice, "without pictures or conversations?"

So she was considering in her own mind (as well as she could, for the hot day made her feel very sleepy and stupid) whether the pleasure of making a daisy-chain would be worth the trouble of getting up and picking the daisies, when suddenly a White Rabbit with pink eyes ran close by her.

There was nothing so *very* remarkable in that; nor did Alice think it so *very* much out of the way to hear the Rabbit say to itself, "Oh dear! Oh dear! I shall be too late!" (when she thought it over afterwards, it occurred to her that she ought to have wondered at this, but at the time it all seemed quite natural); but when the Rabbit actually *took a watch out of its waistcoat-pocket*, and looked at it, and then hurried on, Alice started to her feet, for it flashed across her mind that she had never before seen a rabbit with either a waistcoat-pocket, or a watch to take out of it, and burning with curiosity, she ran across the field after it, and was just in time to see it pop down a large rabbit-hole under the hedge.

2.3. Worksheet 2b: The Meaning of 'Exception'

The word "remarkable" and the expression "out of the way" indicate that something is exceptional. Discussion of the concept "exception" in logic kicks off with Worksheet 2b, which begins with the excerpt shown above (Section 2.2).

Remarks:

- Worksheet 2b is to be completed in pairs or in small groups to allow for consultation.
- It is suggested that upon completion of the worksheet, students present their answers to the class, in order to initiate a whole class discussion of the solutions.
- At this point it is advised to avoid any judgment of the answers; points of disagreement and misperceptions, however, should be noted.
- After presentation of the logical and mathematical background by the teacher (Section 2.6 below), these answers should be revisited and discussed once again.

Worksheet 2b and Proposed Solutions

1. Formulate the rule that caused Alice not to notice anything exceptional when she heard the Rabbit speaking. (It may be helpful to begin your formulation with the word ALL.)

   When Alice discovered a Rabbit speaking, "there was nothing so very remarkable in that; nor did Alice think it so very much out of the way," apparently because it was consistent with the rule: This creature comes from a place where ALL rabbits (and perhaps even all other animals) can speak.

2. When Alice thought it over afterwards, it seemed to her that it should have surprised her that she had heard the Rabbit speaking. What is the rule that drove Alice to think that there was something unusual about it after all?

   The rule that drove Alice when she thought it over afterwards was apparently: ALL the rabbits that I have met until today do not speak. She therefore should have been surprised that in Wonderland there is a talking rabbit.

3. What is the exception that refutes the rule: All prime numbers are odd?

   The number 2 is prime and even; therefore 2 is the exception that refutes the rule. In other words, not all prime numbers are odd since there exists a prime number that is not odd.

4. Is there an exception that refutes the rule: All quadrilaterals in the plane, have the same sum of interior angles, and this sum is 360°? (Note the following mathematical conventions: a quadrilateral is a polygon with four vertices and four sides. A polygon is a finite collection of points in the plane that are ordered in a loop; that is, the first and the last points coincide, and every two adjacent points are connected by a line segment. In short — a polygon is a closed chain of line segments in the plane. The points are the vertices of the polygon and the segments are the sides of the polygon.)

   Consider the quadrilateral ABCD shown in the diagram with sides: AB, BC, CD and DA. The sum of the interior angles of this quadrilateral, namely the acute angles at A, B, C and D is less than 360°. This quadrilateral refutes the rule.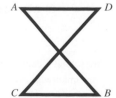

Details on this topic appear in Section 2.6 part C.[1]

5.   Is there an exception that refutes the rule: All triangles in the plane, have the same sum of interior angles, and the sum is 180°?

No. There is no such exception. In every triangle in the plane, the sum of the interior angles is 180°.[2]

6.   In each row presented below there are four cards. Identify the exception among the four cards relative to the remaining three. This may be done by finding a characteristic common to three of the cards that does not exist in the fourth.

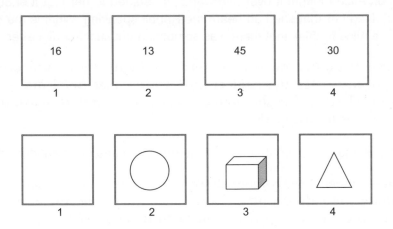

In each case we present one of the many possibilities for which a particular card is exceptional.

In the first row:

Card 1 is exceptional because it is the only one for which the two-digit number on its face is a square number.

Card 2 is exceptional because it is the only one for which the two-digit number on its face is prime.

---

[1] Additional details appear in the following links: http://mathworld.wolfram.com/Polygon.html, http://mathworld.wolfram.com/Quadrilateral.html.

[2] Proof of this theorem appears in the following link: http://mathworld.wolfram.com/Triangle.html.

Card 3 is exceptional because it is the only one for which the two digits of the two-digit number on its face are consecutive.

Card 4 is exceptional because it is the only one for which the sum of the digits of the two-digit number on its face is less than 4. (In all the others the sum of the digits is at least 4.)

In the second row:

Card 1 is exceptional because it is the only one with no shape on its face.

Card 2 is exceptional because the shape on its face is not made up of line segments.

Card 3 is exceptional because the shape on its face represents a three-dimensional shape.

Card 4 is exceptional because the shape on its face is a polygon.

2.4. A Look Inside the Text

Read the first paragraph on page 1, which appears in boldface print below, once again. This paragraph will serve as the basis for discussion of the logical connective NOT.

Down the Rabbit-Hole

 LICE was beginning to get very tired of sitting by her sister on the bank, and of having nothing to do: once or twice she had peeped into the book her sister was reading, but it had no pictures or conversations in it, "and what is the use of a book," thought Alice, "without pictures or conversations?"

So she was considering in her own mind (as well as she could, for the hot day made her feel very sleepy and stupid) whether the pleasure of making a daisy-chain would be worth the trouble of getting up and picking the daisies, when suddenly a White Rabbit with pink eyes ran close by her.

There was nothing so *very* remarkable in that; nor did Alice think it so *very* much out of the way to hear the Rabbit say to itself, "Oh dear! Oh dear! I shall be too late!" (when she thought it over afterwards, it occurred to her that she ought to have wondered at this, but at the time it all seemed quite natural); but when the Rabbit actually *took a watch out of its waistcoat-pocket*, and looked at it, and then hurried on, Alice started to her feet, for it flashed across her mind that she had never before seen a rabbit with either a waistcoat-pocket, or a watch to take out of it, and burning with curiosity, she ran across the field after it, and was just in time to see it pop down a large rabbit-hole under the hedge.

2.5. Worksheet 2c: The Logical Connective NOT

The central topic of this chapter — the logical connective NOT — kicks off with Worksheet 2c, which begins with the excerpt shown above (Section 2.4).
We will discuss this excerpt once again in Worksheet 4c of Chapter 4, following the presentation of De Morgan's laws.

Remarks:

- Worksheet 2c is to be completed in pairs or in small groups to allow for consultation.
- It is recommended that upon completion of the worksheet, students present their answers to the class in order to initiate a whole class discussion of the solutions.
- At this point it is advised that judgment of students' answers be avoided; points of disagreement and misperceptions, however, should be noted.
- After presentation of the logical and mathematical background (Section 2.6 below), these answers should be revisited and discussed once again.

Worksheet 2c and Proposed Solutions

1. Did Alice think there were pictures in the book her sister was reading? Explain your answer.

   No, Alice peeped into the book and there were no pictures. That is why she thought the book had no pictures in it at all.

2. Did Alice think there were conversations in the book her sister was reading? Explain your answer.

   No, Alice peeped into the book and there were no conversations. That is why she though the book had no conversations in it at all.

3. What did Alice see when she peeped into the book her sister was reading? Explain your answer.

   We have no explicit information as to what Alice did see. We may assume that she saw text describing what is happening (but not conversation) and page numbers (but no pictures).

2.6. Logical and Mathematical Background

Upon completion of Worksheets 2b and 2c, and after discussion of the students' answers, the following concepts and topics are to be presented:

    A. The logical connective NOT
    B. The use of the connective NOT in spoken language
    C. The mathematical meaning of 'exception'

A. The logical connective NOT

By adding the word **NOT** to a simple statement, a new statement is obtained; for example: "John did not come to school today" or "Seven is not a prime number." Thus, by convention, **NOT** is referred to as a **logical connective**. A statement that contains the logical connective NOT is no longer considered a simple statement.[3]

Like any other statement, a statement that contains the logical connective NOT may be a TRUE statement or a FALSE statement. Thus, for example, the statement "Seven is not a prime number" is a FALSE statement, while "Eight is not a prime number" is a TRUE statement, negating the simple statement "Eight is a prime number," which itself is a FALSE statement.

The Logical Connective NOT is a **unary logical operator**. The term "logical operator" indicates that its operand is a statement, while "unary" indicates that it operates on a single statement, deriving a new statement from it. If the original statement is a TRUE statement then the new statement will be a FALSE one, and conversely — if the original statement is FALSE then the new statement will be TRUE.

By convention, the symbol for negation of a statement is $\sim p$ (sometimes an alternative symbol is used, such as $\neg p$ or $-p$).

---

[3] In spoken language negation may sometimes be expressed using words other than "not", which turn a statement back into a simple one. For example the negation of "allowed" is "forbidden"; the negation of "with" is "without;" In mathematics too, instead of saying: "The number is 7 not composite" we say: "The number 7 is prime", etc.

Note: Negating the negation of a statement produces the original statement:

$\sim(\sim p) \equiv p$. The symbol '$\equiv$' denotes that the statements are equivalent (equivalence of statements is addressed in Chapter 4).

Basic rule: A statement and its negation cannot both be TRUE and cannot both be FALSE. If one is TRUE then the other is necessarily FALSE (this rule is called "The Law of Excluded Middle").

Truth table for the logical connective NOT

p	$\sim p$
T	F
F	T

Deduction from statement negation:

The negation of a statement does not necessarily indicate what in fact **is** happening. Thus, for example, from the statement "Yesterday it did not snow in the mountains" one cannot deduce that yesterday was a nice day in the mountains. It is possible, for example, that it rained in the mountains yesterday, or perhaps it hailed; it is also possible that the entire day, or a portion of it, had good weather. One cannot deduce the stated conclusion since there are more than two possible states of weather each day. Another example is the negation of the statement "ABC is an equilateral triangle." From its negation — "ABC is NOT an equilateral triangle" — one cannot deduce what kind of triangle ABC is. It may be isosceles, and it may be scalene. Only in cases where there exist **exactly** two possibilities, may one deduce the other possibility from the negation of the first. For example, from the statement "My mother is not alive" one may deduce that the speaker's mother has died. From the statement "Eight is not a prime number" we obtain that 8 is a composite number.[4] This point illustrates the difference between Questions 1 and 2 in Worksheet 2a.

Further details on this topic appear in part 3 of the following section.

---

[4]Note that in the generalized statement "n is not prime" one cannot deduce that n is composite, because n could be 1, which is neither prime nor composite.

B. The use of the connective NOT in spoken language

1. A question such as "Is Michael going home?" has an answer of YES or NO. Someone who is in doubt will often answer "I don't think so," although their intent is "I think not." Similarly, when indicated in the statement in Section 2.2 "nor did Alice think it so *very* much out of the way," the intent is that "Alice thought it not *very* much out of the way." Another example of linguistic usage related to negation may be seen in the following example: The host asks his guest, "You don't drink coffee, right?" The guest replies, "No." Has the guest intended to negate his host's assertion? No, on the contrary — his intent was to confirm it. The guest does not drink coffee. Strange as it may sound, in spoken language NO sometimes means YES.

2. The negation of the negation of a statement is the original statement. But in day-to-day language we often use double negation to emphasize negation and not to get back to the original statement. Using a double negative for emphasis when only one is necessary is non-standard, but is encountered occasionally in spoken language. Shakespeare makes use of the double negative in *Twelfth Night* for emphasis: "And that no woman has, nor never none shall mistress of it be, save I alone." Double negation may also be used to create an effect of understatement, thus the statement "The price of the car is not insignificant," carries a different nuance from the statement "The price of the car is significant."

3. As indicated above, from a logical perspective a statement and its negation indicate opposites, and thus when the first is TRUE the second is necessarily FALSE, and conversely. Nevertheless, in spoken language opposites often have identical meanings. Thus, for example, the expression "You ain't seen nothin' yet," is identical in meaning to "You haven't seen anything yet" even though it is its logical opposite. The reason for this is that in spoken language there is often a lack of clarity or inconsistency in definitions. Thus paradoxically, the words "nothing" and "anything" have, in certain contexts, the same meaning, even though logically they are opposites.

4. In spoken language there are cases where omitting the negation does not create the opposite meaning. Thus, for example, "It is not possible that it is raining now" is a definitive assertion that at this time it is not raining. On the other hand, "It is possible that it is raining now" is not a definitive assertion that it is raining now, and in this sense it is not the opposite of the previous assertion. This also applies to "it couldn't be that —,"among other examples.

5. A simple statement p is viewed in logic as an **atom**; that is, a unit which cannot be further broken down into parts, and which can be in only one of two states — TRUE or FALSE. The simple statement may be negated, deriving the statement *NOT p*. When we try to apply this to simple statements in spoken language, we encounter a problem. The problem is that when we negate a simple statement in spoken language, the negation can be applied to any part of the statement. For example, if the statement was: "Michael drove to the big city," then the formal negation would be: "Not Michael drove to the big city" meaning — it is not true that Michael drove to the big city. Why is this wrong? Is it because not Michael drove to the big city (but rather someone else)? Or perhaps because Michael did not drive (but rather flew) to the big city? Or it's possible that Michael drove somewhere else and not to the big city? We cannot answer this with the information we have at hand. In spoken language we tend to insert the negation before the verb and say "Michael did not drive to the big city." During the course of our speech we use vocal variations for emphasis in order to indicate the different meanings we wish to lend to the negation. If we place the emphasis on Michael, as in "**Michael** did not drive to the big city," the listener understands that it was not Michael who drove to the big city. On the other hand, if we place the stress as "Michael did not **drive** to the big city," the listener can understand that Michael arrived in the big city not by driving but by some other means. In any event, it is clear that the negation of "Michael drove to the big city" is that it is not true that Michael drove to the big city, whatever the reason.

6. From the logical perspective, negation of a **compound statement** is more complex. For example, negation of the statement "Today it is raining in the mountains and Michael flew to Europe" would be "It is not the case that today it is raining in the mountains and Michael flew to Europe." We will revisit the negation of the complex statement in Chapter 4 when we address De Morgan's Laws.

C. The mathematical meaning of "exception"

The third paragraph of *Alice's Adventures in Wonderland* opens with the words: "There was nothing so *very* remarkable in that; nor did Alice think it so *very* much out of the way to hear the Rabbit say to itself...".

In order to understand this sentence precisely, notice that the first part of it is the negation of the statement: "There was something so *very* remarkable in that...". The subsequent portion is the negation of "Alice thought it so *very* much out of the way...".

We will use this as a springboard to understand the linguistic and mathematical meaning of 'exceptional' or 'anomalous' more deeply. To the reader, and later on to Alice herself, it does seem unusual or exceptional that a Rabbit could speak, as the well-known rule is: "All animals do not speak".[5] This last statement opens with the word ALL, the logical meaning of which will be addressed in detail in Chapter 7.

What is it that makes hearing the Rabbit speak something that is not exceptional? It seems that Alice, from the start, associated rules with Wonderland that don't apply in our world, and thought that in Wonderland all rabbits, and perhaps even all other animals could speak. That is why she was not surprised when she encountered a talking Rabbit.

The key to something being considered an exception is the existence of a rule that applies to a collection of objects, where there exist one or several individual objects of the same type to which the rule does not apply. For example:

- (Almost) every day, the sun rises and shines throughout the morning hours. But there are a few days on which a solar eclipse occurs; in those cases the sun is hidden during the daytime hours and does not shine. A solar eclipse is therefore an exceptional event, unusual, an exception to the rule.

- (Almost) every year, February has 28 days. The exceptions to the rule are the leap years.

It is evident that the word ALL (or EVERY) describes rules.

Note that in spoken language, words associated with exceptions may have a positive or negative connotation. For example, the words "oddity" or "deviant"

---

[5] Later in the story it becomes evident that Alice does in fact see talking animals as an exceptional phenomenon. Alice meets a mouse on page 21 of the book and wonders whether she should address it: "'Would it be of any use, now,' thought Alice, 'to speak to this mouse? Everything is so out-of-the-way down here that I should think very likely it can talk: at any rate, there's no harm in trying.'"

are associated primarily with something undesirable, while "extraordinary" often indicates wonderment. In mathematics the use of these words is non-judgmental.

In mathematics the words ALL or EVERY are used, among other things, **to define concepts**. For example: Every positive whole number that has exactly two distinct divisors (1 and the number itself) is called a **prime number.** Mathematicians go to great lengths to define mathematical concepts so that there are no exceptions. They also seek common characteristics of mathematical objects. So, for example, all prime numbers are odd with the exception of the prime number 2. The number 2 is therefore an exception in the set of prime numbers. We can use the logical connective NOT to indicate this: NOT all prime numbers are odd. In other words, the statement "All prime numbers are odd" is FALSE, because there exists one prime number that is exceptional, being that it is even. This can also be expressed in the statement: "Almost all prime numbers are odd." The meaning of "almost all" is "except for a finite (and relatively small) number of exceptions."

Thus, the exception refutes the truth value of the rule. In mathematics, an exception like this is called a **counterexample.** A counterexample refutes the conjecture about the truth of a rule.

Let us consider the following conjecture: In the Euclidean plane, the sum of the interior angles of all quadrilaterals is a constant (is equal to 360°). Note the following mathematical conventions: (a) A quadrilateral is a polygon with four vertices and four sides. (b) A polygon is a collection of points in the plane that are ordered in a loop; that is, the first and the last points coincide, and every two adjacent points are connected by a line segment. In short — a polygon is a closed chain of line segments in the plane. The points are the vertices of the polygon and the segments are the sides of the polygon.

There exist three types of quadrilaterals: (a) convex (as in the figure on the left); (b) concave (as in the middle figure); (c) crossed (as in the figure on the right). It is obvious that the assertion above may be confirmed with the convex and the concave quadrilaterals, as follows: Each quadrilateral of this type may be divided into two triangles by drawing an interior diagonal (as shown in the figures), such that their interior angles coincide with the interior angles of the quadrilateral. Now, based upon the theorem that for each triangle in the Euclidean plane (without exception!)

the sum of the interior angles is a constant, and is equal to 180°, we obtain that 360° is the sum of the angles of quadrilaterals of these types.

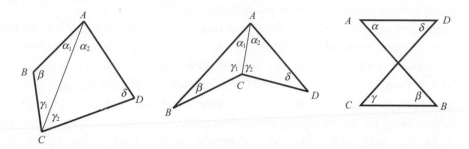

The proof is based on the fact that in convex and concave quadrilaterals *ABCD* it is possible to draw at least one interior diagonal (*AC* or *BD*). The case of crossed quadrilaterals, however, is different. In these, the sum of the interior angles is less than 360° and is not even constant for all quadrilaterals.

In summary, the assumption that for all quadrilaterals in the Euclidean plane the sum of the interior angles is the same has been refuted through the counterexample of crossed quadrilaterals. Nevertheless, it *is* true that the sum of the interior angles of all convex quadrilaterals in the Euclidean plane is constant, and the assertion is similarly true for all concave quadrilaterals. In every such quadrilateral, the sum of the interior angles is a constant, and is equal to 360°. The proof above guarantees that there cannot be a counterexample to this assertion.

2.7. Review of Answers to Worksheets 2b and 2c

After students have become familiar with the logical background, it is advised to ask them to go back and review their answers to Worksheets 2b, 2c and correct them if necessary.

2.8. Worksheet 2d: The Logical Connectives NOT and AND — Inference Using a Truth Table

Worksheet 2d addresses inferences from statements containing the logical connectives NOT and AND by using truth tables.

Remarks:

- This worksheet may be completed as individual practice, in pairs or in small groups, with group discussion encouraged.
- Upon completion of the worksheet, it is recommended that the students present their answers to the class to initiate a whole class discussion of the solutions.
- In answering the questions in this worksheet, the rule that if a statement is TRUE then its negation is FALSE, and conversely, should be kept in mind. Similarly, as we saw in the previous chapter, if two simple statements are linked using the binary logical connective AND, the result is a compound statement the truth value of which depends on the truth value of its two components.

Worksheet 2d and Proposed Solutions

Alice wandered about in the wood of Wonderland in an attempt to find something that would help her grow, as her height at that moment was only three inches. In the course of her wandering Alice met a blue caterpillar sitting on the top of a mushroom and smoking a hookah. The caterpillar asked her who she was, but the confused Alice did not know what to answer him. Here is an excerpt from the conversation between Alice and the caterpillar (page 49).

Alice has a Conversation with the Caterpillar

"What do you mean by that?" said the Caterpillar sternly. "Explain yourself!"

"I can't explain *myself*, I'm afraid, sir," said Alice, "because I'm not myself, you see."

'I don't see,' said the Caterpillar.

"I'm afraid I can't put it more clearly," Alice replied very politely, "for I can't understand it myself to begin with;

and being so many different sizes in a day is very confusing."

"It isn't," said the Caterpillar.

Each item below contains two statements followed by a question. Circle the correct answer and explain your reasoning based on the given information.

1.   (i) In Wonderland the following statement is FALSE:
        "Alice can explain herself, and Alice expresses things clearly."

     (ii) In Wonderland the following statement is TRUE:
        "Alice can explain herself."

     Which of the following is correct?

     A.   The statement "Alice expresses things clearly" is TRUE.

     B.   **The statement "Alice expresses things clearly" is FALSE.**

     C.   The two given statements contradict one another.

     D.   There is insufficient information given to determine how well Alice expresses things.

Reasoning:

Let us designate the statements as follows:

p: Alice can explain herself.

q: Alice expresses things clearly.

Given: The compound statement p∧q is FALSE.

In accordance with the truth table for the logical connective AND, rows 2–4 match the given information.

p	q	$p \wedge q$
T	T	T
T	F	F
F	T	F
F	F	F

It is also given that p is a TRUE statement; thus, now only the second row of the table matches the given information. Therefore we obtain that statement q, that is, "Alice expresses things clearly," is FALSE. This corresponds to option B, which is marked as the correct answer.

p	q	$p \wedge q$
T	T	T
T	F	F
F	T	F
F	F	F

The question may also be answered without using a truth table. Since the compound statement p∧q is FALSE, at least one of its two components — the simple statements p or q — is FALSE. Because p is TRUE, we conclude that q must be FALSE.

2. The following two statements are TRUE in Wonderland:

(i)   The caterpillar does not understand, and Alice is very short.
(ii) Alice is very short.

Which of the following is correct?

A.   The statement "The caterpillar understands" is TRUE.

B.   **The statement "The caterpillar understands" is FALSE.**

C.   The two given statements contradict one another.

D.   There is insufficient information given to determine whether the caterpillar understands.

Reasoning:

Let us designate the statements as follows:

p: The caterpillar does not understand.

q: Alice is very short.

Given: The compound statement p∧q is TRUE.

p	q	$p \wedge q$
T	T	T
T	F	F
F	T	F
F	F	F

In accordance with the truth table for the logical connective AND, the first row matches the given information.

It is also given that q is a TRUE statement; therefore p is also a TRUE statement. That is, "The caterpillar does not understand" is a TRUE statement. Therefore, we obtain that "The caterpillar understands" is FALSE. This corresponds to option B, which is marked as the correct answer.

The question may also be answered without using a truth table. Since the compound statement p∧q is TRUE, both of its components — the two simple statements — must be TRUE as well. Given that q is TRUE, it follows that p must be TRUE as well. That is, ~p must be FALSE.

3.   Both of the following statements are FALSE in Wonderland:

    (i)   Alice's height changed during the day, and the Caterpillar understands Alice.

    (ii)   The Caterpillar understands Alice.

Which of the following is correct?

A.   The statement "Alice's height did not change during the day" is TRUE.

B.   The statement "Alice's height did not change during the day" is FALSE.

C.   The two given statements contradict one another.

D.   **There is insufficient information given to determine what happened to Alice's height during the day.**

Reasoning:

Let us designate the statements as follows:

p: Alice's height changed during the day.

q: The caterpillar understands Alice.

Given: The compound statement p∧q is FALSE.

p	q	$p \wedge q$
T	T	T
T	F	F
F	T	F
F	F	F

In accordance with the truth table for the logical connective AND, rows 2–4 match the given information.

It is also given that q is FALSE. Thus we are left with rows 2 and 4 of the table. Hence p can be either TRUE or FALSE. Therefore, the given information does not suffice to answer the question. This corresponds to option D, which is marked as the correct answer.

p	q	$p \wedge q$
T	T	T
T	F	F
F	T	F
F	F	F

The question may also be answered without using a truth table. Since the compound statement p∧q is FALSE, at least one of its two components — the simple statements p or q — is FALSE. Since q is FALSE, we cannot determine whether p is TRUE, or whether it is FALSE.

4. Both of the following statements are TRUE in Wonderland:

    (i) Alice feels very strange, and Alice cannot explain herself.
    (ii) Alice can explain herself.

    Which of the following is correct?

    A. The statement "Alice feels very strange" is TRUE.

    B. The statement "Alice feels very strange" is FALSE.

    C. **The two given statements contradict one another.**

    D. There is insufficient information given to determine whether Alice feels strange.

Reasoning:

Let us designate the statements as follows:

p: Alice feels very strange.

q: Alice cannot explain herself.

Given: The compound statement p∧q is TRUE.

p	*q*	*p ∧ q*
T	T	T
T	F	F
F	T	F
F	F	F

In accordance with the truth table of the logical connective AND, the first row matches the given information.

It is also given that the statement "Alice can explain herself" is TRUE. This statement is the negation of statement q; therefore statement q is FALSE. Thus, rows 2 and 4 match the given information.

p	*q*	*p ∧ q*
T	T	T
T	F	F
F	T	F
F	F	F

Since we did not find rows matching the two truth tables, our conclusion is that the two given data items contradict one another. This corresponds to option C, which is marked as the correct answer.

The question may also be answered without using a truth table. Since the compound statement p∧q is TRUE, our conclusion is that its two components — the simple statements p and q — are necessarily TRUE. Since ~q is TRUE, our conclusion is that q is FALSE. Therefore the given information is contradictory.

2.9. Worksheets 2e and 2f: Summary Exercise for the Logical Connective NOT

The summary exercise is composed of two Worksheets: 2e and 2f.

Remarks:

- Worksheet 2e is to be completed as individual practice by students. In Worksheet 2f (which appears in the students' workbook only) students will compare their current answers with the answers they gave at the start of the chapter (Worksheet 2a).
- Upon completion of the summary exercise, it is recommended that a whole class discussion will be held to consider the changes that have taken place in students' understanding and perceptions through the course of this chapter, as well as to identify the particular difficulties encountered with the subject matter.

Worksheet 2e and Proposed Solutions

Answer the questions in this worksheet in their entirety, providing as much detail as possible. If necessary, you may indicate: "I did not understand the question, therefore I have not answered it." Make sure not to go back to Worksheet 2a before you complete your work on this worksheet.

1. The following statement contains a negation: "Lewis Carroll, writer and mathematician, is no longer alive." Can this statement be rephrased without using the word NO (or an equivalent such as NOT) without changing the original meaning? If so, how? If not, why not?

   The statement "Lewis Carroll, writer and mathematician, is no longer alive" can be rephrased as follows: "Lewis Carroll, writer and mathematician, has already died." The ability to do so stems from the fact that the negation of "not alive" is "dead" — there is no other negation to "not alive." This condition is referred to as dichotomous.

2. The following statement also contains a negation: "The White Rabbit did not eat the carrot." Can this statement be rephrased without using the word NO (or an equivalent such as NOT) without changing the original meaning? If so, how? If not, why not?

   The statement: "The White Rabbit did not eat the carrot" cannot be rephrased without using the word NO or NOT, since it is not clear what the White Rabbit did in fact do. "...did not eat the carrot" may be interpreted in several ways, and none of these is the equivalent of the original statement, for example: "The White Rabbit ate a tomato," "The White Rabbit threw the carrot," etc. This is because contrary to the case in Question 1, the condition in this case is not dichotomous — there exist more than one possible negation for the original statement.

3. The following statement contains a double negative: "It is not possible that Alice does not like daisies." Can this statement be rephrased without using the word NO (or the equivalent such as NOT) without changing the original meaning? If so, how? If not, why not?

   The statement: "It is not possible that Alice does not like daisies" can be rephrased as: "Alice likes daisies." The presence of a double negative in a statement gives the statement a positive meaning.

4. What is the meaning of the word NO? If it has more than one meaning, indicate all meanings and provide examples.

    The word NO negates a statement and changes its truth value from TRUE to FALSE, and conversely. Similarly, in spoken language, the word NO is often used for emphasis, and in such a case a double negative may not necessarily provide the intended logical meaning.

Chapter 3

The Logical Connective OR

Who is the book about —
Alice OR the Rabbit?

Who draws the curtain —
Alice OR the Rabbit?

3.1. Worksheet 3a: Opening Exercise. Lead-in to the Logical Connective OR

Instruction of this chapter begins with Worksheet 3a: Opening Exercise (hereforth abb. WS3a) to be completed as individual practice by each student. Upon completion of the task, notify the students that they will return to this worksheet at the end of the chapter (Worksheet 3h Summary Exercise); then they will be able to compare their initial understanding with that which they acquired during the course of the chapter, while reflecting upon the changes in their understanding along with the points where further clarification is still required (Worksheet 3i, which appears only in the Student's Workbook). For this reason it is recommended that discussion of this worksheet will be held off until the end of the chapter. In some cases the teacher may prefer to collect WS3a and give it back only upon completion of WS3h when students are ready to work on WS3i.

Because this lead-in worksheet is repeated again as a summary exercise, the sheet (together with proposed solutions) appears only once — at the end of this chapter.

3.2. A Look Inside the Text

Read the portion of the third paragraph of *Alice's Adventures in Wonderland* on page 1 that appears in boldface. These lines will serve as the basis for discussion of the logical connective OR.

Down the Rabbit-Hole

 LICE was beginning to get very tired of sitting by her sister on the bank, and of having nothing to do: once or twice she had peeped into the book her sister was reading, but it had no pictures or conversations in it, "and what is the use of a book," thought Alice, "without pictures or conversations?"

So she was considering in her own mind (as well as she could, for the hot day made her feel very sleepy and stupid) whether the pleasure of making a daisy-chain would be worth the trouble of getting up and picking the daisies, when suddenly a White Rabbit with pink eyes ran close by her.

There was nothing so *very* remarkable in that; nor did Alice think it so *very* much out of the way to hear the Rabbit say to itself, "Oh dear! Oh dear!

I shall be too late!" (when she thought it over afterwards, it occurred to her that she ought to have wondered at this, but at the time it all seemed quite natural); **but when the Rabbit actually *took a watch out of its waistcoat-pocket*, and looked at it, and then hurried on, Alice started to her feet, for it flashed across her mind that she had never before seen a rabbit with either a waistcoat-pocket, or a watch to take out of it, and burning with curiosity, she ran across the field after it, and was just in time to see it pop down a large rabbit-hole under the hedge.**

3.3. Worksheet 3b: The Logical Connective OR

The central topic of this chapter — the logical connective OR — kicks off with Worksheet 3b, which begins with the excerpt shown above (Section 3.2).

Remarks:

- Worksheet 3b is to be completed by students in pairs or in small groups to allow for consultation.
- It is suggested that upon completion of the worksheet, students present their answers to the class, in order to initiate a whole class discussion of the solutions.
- At this point it is advised to avoid any judgment of students' answers; points of disagreement and misperceptions, however, should be noted.
- After presentation of the logical and mathematical background by the teacher (Section 3.7 below), students' answers should be revisited and discussed once again.

Worksheet 3b and Proposed Solutions

1. Did the Rabbit have a watch? How do you know that?

   Yes, it had a watch; we know this because it was able to take the watch out of its waistcoat-pocket.

2. Did the Rabbit have a waistcoat-pocket? How do you know that?

   Yes, it had a waistcoat pocket, because it says that the Rabbit took the watch out of it.

3. Is the statement: "The Rabbit has a waistcoat-pocket and a watch" correct? Why?

   Yes; this may be derived from the answers to Questions 1 and 2.

4. Can we replace the sentence: "… she had never before seen a Rabbit with either a waistcoat-pocket, or a watch to take out of it…" with the sentence: "… she had never before seen a Rabbit with a waistcoat-pocket, and a watch to take out of it…" without changing the meaning of the original? Why?

   Yes; in this case both sentences have identical meaning. The logical connective OR always includes the AND case, unless the content-related context contradicts that possibility. Nothing in our original sentence negates the possibility that both propositions occur; moreover, it is stated explicitly in the previous sentence that both did in fact occur.

5. If your answer to the previous question was in the affirmative, then do you think that this implies that OR and AND have identical meaning?

   No. OR and AND do not mean the same thing. The meaning of OR includes the meaning of AND, but also includes the possibility that only one of the two propositions will occur ("Inclusive OR," details on this appear later in sections 3.5, 3.6 and 3.7).

3.4. Another Look Inside the Text

Continue reading on pages 2 and 3 of *Alice's Adventures in Wonderland*, the parts that appear in boldface below. The last sentence in bold will serve as the basis for the discussion on the logical connective OR.

> **In another moment down went Alice after it, never once considering how in the world she was to get out again.**
>
> **The rabbit-hole went straight on like a tunnel for some way, and then dipped suddenly down, so suddenly that Alice had not a moment to think about stopping herself before she found herself falling down what seemed to be a very deep well.**
>
> **Either the well was very deep, or she fell very slowly, for she had plenty of time as she went down to look about her, and to wonder what was going to happen next.** First, she tried to look down and make out what she was coming to, but it was too dark to see anything; then she looked at the sides of the well and noticed that they were filled with cupboards and book-shelves: here and there she saw maps and pictures hung upon pegs. She took down a jar from one of the shelves as she passed; it was labelled "ORANGE MARMALADE," but to her disappointment it was empty.

3.5. Worksheet 3c: The Difference Between Exclusive OR and Inclusive OR

The objective of Worksheet 3c, which begins with the passage mentioned above (Section 3.4), is to underscore the distinction between OR (sometimes referred to as Inclusive OR) and Exclusive OR.

Remarks:

- Worksheet 3c is to be completed by students in pairs or in small groups to allow for consultation.
- It is suggested that upon completion of the worksheet, students present their answers to the class in order to initiate a whole class discussion of the solutions.
- At this point it is advised to avoid any judgment of students' answers; points of disagreement and misperceptions, however, should be noted.
- After presentation of the logical and mathematical background by the teacher (Section 3.7 below), students' answers should be revisited and discussed once again.

Worksheet 3c and Proposed Solutions

1. If we assume that the well was in fact very deep, can we infer from the passage that Alice did not fall very slowly?

    A. Yes, with certainly.

    B. Absolutely not.

    (C.) **It cannot be determined — maybe yes and maybe no.**

    Circle the best answer and give your reasoning:

    It cannot be determined because the logical connective OR includes the possibility of AND, unless that option is ruled out based on the context of the story.

2. If we assume that Alice did in fact fall very slowly, can we infer from the passage that the well was not very deep?

    A. Yes, with certainty.

    B. Absolutely not.

    (C.) **It cannot be determined — maybe yes and maybe no.**

    Circle the best answer and give your reasoning:

    It cannot be determined because the logical connective OR includes the possibility of AND, unless that option is ruled out based on the context of the story.

3. Is there a difference between the meaning of OR in the passage that appears in Worksheet 3b and that which appears in the passage on this worksheet? Explain your answer.

    In Worksheet 3b it was clear from the context of the story that both options connected by OR were TRUE with certainty. In the current worksheet it is not clear from the context of the story whether both options connected by OR are TRUE. It is possible that only one of them is TRUE.

3.6. Worksheet 3d: More on the Difference Between Exclusive OR and Inclusive OR

Worksheet 3d has two objectives: it continues with the presentation of the central topic of this chapter — the logical connective OR, and it underscores the difference between the Exclusive OR and the Inclusive OR.

Worksheet 3d and Proposed Solutions

Alice wandered about Wonderland somewhat confused, after having met some of the creatures who live there.

Alice was 3 inches tall when she met a caterpillar sitting on top of a mushroom and smoking a hookah. The caterpillar tried, in its own particular way, to befriend Alice. The conversation between them did not proceed amicably, and the caterpillar took offense to what Alice said. The caterpillar put its hookah into its mouth and began smoking again (page 55).

Alice Attempts to Talk to the Caterpillar Once Again

This time Alice waited patiently until it chose to speak again.
In a minute or two the Caterpillar took the hookah out of its mouth and yawned once or twice, and shook itself.
Then it got down off the mushroom, and crawled away into the grass.

The word OR appears in this passage twice.

Are there differences between the OR that appears in Worksheet 3b, the OR that appears in Worksheet 3c and the OR that appears on this worksheet? Explain your answer.

In Worksheet 3b it was clear from the context of the story that both options connected by OR were TRUE with certainty. In Worksheet 3c it was not clear from the context whether the options connected by the OR were TRUE. It is possible that only one of the two was in fact TRUE.

In this worksheet it is clear from the context that only one of the two options connected by the OR is TRUE. If the caterpillar was silent for exactly one minute, then it is not possible that it was silent for two minutes; conversely, if it was silent for two minutes then it is not possible that it was silent for one minute. Similarly, if the caterpillar yawned only once, it could not possibly have yawned twice; and conversely. That is, in the two cases before us, if one is TRUE then it is clear that the other is necessarily FALSE. In other words — it is impossible for the two options to exist simultaneously.

3.7. Logical and Mathematical Background

Upon completion of Worksheets 3b, 3c and 3d, and after discussion of students' answers, the following concepts and topics are to be presented:

A. The use of the logical connective OR to obtain a compound statement
B. Remarks on the meaning of OR

A. The use of the logical connective OR to obtain a compound statement

In Chapter 1 we addressed the formation of a compound statement from two simple statements or their negation through the use of the logical connective AND. In this section we will focus on forming a compound statement from two simple statements by using the **logical connective OR**, also a **binary logical operator**.

Thus, for example, from the unconnected simple statements: "Today it is raining" and "Bob is going to the movies," the compound statement "Today it is raining or Bob is going to the movies" may be obtained. Let us assume for a moment that the first of the two simple statements is TRUE. The compound statement is TRUE whether or not Bob is going to the movies; that is, whether "Bob is going to the movies" is a TRUE statement or a FALSE statement.

As a rule, if two simple statements are linked using the binary logical connective OR, the result is a compound statement whose truth value depends upon the truth value of its two component simple statements.

Its truth value is TRUE if the truth value of **at least one** of its two component statements is TRUE. If the truth value of both of its component statements is FALSE, then the compound statement has a FALSE truth value.

We now return to our example. In addition to what was stated above, the compound statement is TRUE even if it is not raining, on the condition that Bob is in fact going to the movies.

Needless to say, the compound statement is TRUE if both of its components are TRUE; that is, if both "Today it is raining" and "Bob is going to the movies" are TRUE. There is a possibility that someone may interpret the compound statement as stating that if it rains, Bob will not go to the movies. But this interpretation is incorrect. Just because one of the options is true, it does not exclude the possibility that the other option is true as well.

It follows that OR always includes the possibility of AND (since if both components are TRUE, this is a case of AND), unless this option is not a reasonable one. An example of this is the compound statement obtained from two simple statements: "It is raining here now" and "It is not raining here now." The compound statement obtained from these would be: "It is raining here now or it is not raining here now." Clearly this statement is TRUE, because one of the two options is always TRUE, but both options will never occur simultaneously (we will expand upon this in the next section).

The conventional notation for the statement p OR q is $p \vee q$.

The source of the symbol $\vee$ is the first letter of the Latin word *vel*, which means OR.

Truth Table for the Logical Connective INCLUSIVE OR

p	q	$p \vee q$
T	T	T
T	F	T
F	T	T
F	F	F

B. Remarks on the meaning of OR

As noted, OR always includes the option of AND unless such a possibility is not reasonable (as in the example given above). The word OR is used on occasions when it is not reasonable that both options will exist simultaneously, or when they contradict one another. An example of this is the statement "Chris is taking AP math or (Chris is taking) remedial math," since each student can be in only one of the two classes — either the AP math class OR the remedial math class — but not in both. In cases like these, OR is sometimes referred to as **Exclusive OR**, since the AND option is excluded.

The word OR is sometimes used in the sense of Exclusive OR even when there is no contradiction between the two options. For example, Joseph tells Lisa about Richard's birthday presents, and says: "Richard got a German Shepherd or (Richard got) a Labrador." Perhaps Joseph wants to say that he doesn't remember exactly what Richard told him, but he is sure that Richard got a dog (Exclusive OR). Nevertheless, it is also possible that Joseph wanted to say that he is not sure whether Richard got one of the two kinds of dogs, or whether he got both a German Shepherd and a Labrador (**Inclusive OR**).

Use of Exclusive OR is found in particular in language that has some degree of threat. For example, the doctor orders her patient: "At breakfast, you may eat only one fruit — an apple or an orange." From the words "only one" it is clear that the doctor's intention was Exclusive OR, and that the patient may not eat both an apple and an orange. Had the doctor not stressed that, one may have understood from "you may eat an apple or (you may eat) an orange" that the patient can eat both an apple and an orange. A similar case occurs when a parent says to a child: "Finish your homework or you will not watch TV." From the context it is clear that the intent is not that the child will finish his homework and then the parent will not let him watch TV. We often stress the special meaning of Exclusive OR by using "either... or else…".

In general, use of OR requires contextual consideration. Nonetheless, where context does not indicate otherwise, and in mathematics in particular, OR refers to INCLUSIVE OR.

3.8. Review of Answers to Worksheets 3b, 3c and 3d

After students have become familiar with the logical background, it is advised to ask them to go back and review their answers to Worksheets 3b, 3c and 3d and correct them if necessary.

3.9. Worksheet 3e: Summary Exercise for the Difference Between (Inclusive) OR and Exclusive OR

The objective of Worksheet 3e is to allow students to formulate the differences between OR (Inclusive OR) and Exclusive OR after having learned the distinctions between them. The activity is to be carried out as a group exercise. The students will be asked to describe an imaginary dialogue between Alice and the White Rabbit, during which the White Rabbit tries to explain to the befuddled Alice the differences between the two kinds of OR connectives.

Worksheet 3e

Group task:

Alice, being a very curious girl, tends to ask the White Rabbit many questions in order to gain a profound understand of everything. The White Rabbit, on its part, does its best to try to explain things to Alice through the use of examples and of generalization. Write an imaginary dialogue between the White Rabbit and Alice in which the Rabbit tries to explain to Alice the difference between Exclusive OR and (Inclusive) OR.

3.10. Worksheet 3f: The Logical Connectives OR, AND and NOT — Inference Using a Truth Table

Worksheets 3f and 3g address inferences from statements containing the logical connectives OR, AND and NOT by using truth tables.

Remarks:

- This worksheet may be completed as individual practice, in pairs or in small groups, with group discussion encouraged.
- Upon completion of the worksheet, it is recommended that the students present their answers to the class to initiate a whole class discussion of the solutions.

Worksheet 3f and Proposed Solutions

One summer day the Queen of Hearts, despotic ruler of Wonderland, made some tarts. Someone ate all the tarts, and the Queen suspected the Knave of Hearts. The Queen decided to put the Knave of Hearts on trial. The hot-tempered King was the judge, while the White Rabbit was the herald. There were twelve jurors as well. After the indictment was read, the King commanded the jury to consider their verdict. But the White Rabbit reminded the King that the trial must be conducted in accordance with the rules. The King asked to call as witness the Hatter, a resident of Wonderland in the millinery trade. The Hatter was nervous and was therefore unable to give his evidence. The King became angry and threatened the Hatter. The Hatter could not calm down, and the King therefore threatened him once again (page 145).

"Give your evidence,"
the King repeated angrily,
"or I'll have you executed,
whether you're nervous or not."

Each item below contains two statements followed by a question.

Circle the correct answer and explain your reasoning based on the given information.

1.   Both of the following statements are TRUE in Wonderland:

    (i)   The Hatter will give his evidence or he will be executed.
    (ii) The Hatter will give his evidence.

    Which of the following is correct?

    A.   The statement "The Hatter will be executed" is TRUE.

    B.   The statement "The Hatter will be executed" is FALSE.

C.   The two given statements contradict one another.

(D.) **There is insufficient information given to determine whether the Hatter will be executed or not.**

Your reasoning:

Let us designate the statements as follows:

p: The Hatter will give his evidence.

q: The Hatter will be executed.

Given: The compound statement p ∨ q is TRUE.

p	q	$p \vee q$
T	T	T
T	F	T
F	T	T
F	F	F

In accordance with the truth table for the logical connective OR, rows 1-3 match the given information.

It is also given that p is a TRUE statement. Thus we are left with rows 1 and 2 of the table matching the given information. Hence we conclude that statement q can be either TRUE or FALSE. That is, the given information does not suffice to answer the question regarding the fate of the Hatter. This corresponds to option D, which is marked as the correct answer.

p	q	$p \vee q$
T	T	T
T	F	T
F	T	T
F	F	F

Note: The answer above relies on the assumption that the use of OR in the first statement is the conventional one; namely, the Inclusive OR. Nevertheless, note the use of threatening language in the first statement; ordinarily we would expect that the Hatter would not be executed if he gives his evidence. In Wonderland, however, anything is possible... It is possible that the Hatter will give his testimony and still be executed. And, in fact, as the story proceeds we learn that when the King permitted the Hatter to leave the courtroom after he gave his evidence, the Queen turned to one of the officers of the court and said: "...and just take his head off outside" (page 148).

2.   Both of the following statements are TRUE in Wonderland:

(i)   The Hatter will not give his evidence or the Hatter will not be executed.
(ii) The Hatter will give his evidence.

Which of the following is correct?

A.   The statement "The Hatter will be executed" is TRUE.

B.   **The statement "The Hatter will be executed" is FALSE.**

C.   The two given statements contradict one another.

D.   There is insufficient information given to determine whether the Hatter will be executed or not.

Your reasoning:

Let us designate the statements as follows:

p: The Hatter will not give his evidence.

q: The Hatter will not be executed.

Given: The compound statement p ∨ q is TRUE.

p	*q*	*p ∨ q*
T	*T*	*T*
T	*F*	*T*
F	*T*	*T*
F	*F*	*F*

In accordance with the truth table for the logical connective OR, rows 1–3 match the given information.

It is also given that ~p is a TRUE statement, that is, p must be FALSE. Thus we are left with only row 3 of the table.

p	*q*	*p ∨ q*
T	*T*	*T*
T	*F*	*T*
F	*T*	*T*
F	*F*	*F*

Our conclusion is that statement q is a TRUE statement. That is, "The Hatter will not be executed" is TRUE. Therefore the statement: "The Hatter will be executed" is FALSE. This corresponds to option B, which is marked as the correct answer.

3. Both of the following statements are TRUE in Wonderland:

    (i)   The Hatter will give his evidence and the Hatter is nervous.
    (ii) The Hatter will give his evidence.

    Which of the following is correct?

    A.   The statement "The Hatter is not nervous" is TRUE.

    B.   **The statement "The Hatter is not nervous" is FALSE.**

    C.   The two given statements contradict one another.

    D.   There is insufficient information given to draw conclusions as to whether or not the Hatter is nervous.

    Your reasoning:

    Let us designate the statements as follows:

    p: The Hatter will give his testimony.

    q: The Hatter is nervous.

    Given: The compound statement $p \wedge q$ is TRUE.

    In accordance with the truth table for the logical connective AND, the first row matches the given information.

    It is also given that p is a TRUE statement. Therefore we obtain that the statement q must be TRUE; that is, the statement "The Hatter is nervous" is TRUE. Thus, the statement "The Hatter is not nervous" is FALSE.

p	q	$p \wedge q$
T	T	T
T	F	F
F	T	F
F	F	F

    This corresponds to option B, which is marked as the correct answer.

4. Both of the following statements are FALSE in Wonderland:

    (i)   The Hatter will give his evidence or the Hatter will not be executed.
    (ii) The Hatter will not give his evidence.

Which of the following is correct?

A.   The statement "The Hatter will be executed" is TRUE.

B.   The statement "The Hatter will be executed" is FALSE.

C.   **The two given statements contradict one another.**

D.   There is insufficient information given to draw conclusions regarding the
        Hatter's fate.

Your reasoning:

Let us designate the statements as follows:

p: The Hatter will give his evidence.

q: The Hatter will not be executed.

*Given: The compound statement p ∨ q is
FALSE.*

p	q	$p \vee q$
T	T	T
T	F	T
F	T	T
F	F	F

*In accordance with the truth table of the
logical connective OR, the fourth row matches the given
information.*

*It is also given that ~p is FALSE. That is, p is a TRUE statement.
Therefore we conclude that the two statements contradict one
another. This corresponds to option C, which is marked as the
correct answer.*

5.   In Wonderland the following statement is FALSE:
      "The Hatter is not nervous or the Hatter will be executed."

      In Wonderland the following statement is TRUE:
      "The Hatter will give his evidence."

      Which of the following are correct (there may be more than one right answer)?

      A.   The statement "The Hatter is not nervous and [the Hatter] will give his
              evidence" is TRUE.

B. **The statement "The Hatter is not nervous and [the Hatter] will give his evidence" is FALSE.**

C. **The statement "The Hatter is not nervous or the Hatter will give his evidence" is TRUE.**

D. The statement "The Hatter is not nervous or the Hatter will give his evidence" is FALSE.

E. The two given statements contradict one another.

F. There is insufficient information given to choose one (or more) of the above options.

Your reasoning:

Let us designate the statements as follows:

p: The Hatter is not nervous.

q: The Hatter will be executed.

r: The Hatter will give his evidence.

p	q	$p \vee q$
T	T	T
T	F	T
F	T	T
F	F	F

Given: The compound statement $p \vee q$ is FALSE.

In accordance with the truth table of the logical connective OR, the fourth row matches the given information. We deduce that statement p is FALSE, and also that statement q is FALSE.

Let us now examine whether the statement $p \wedge r$ is TRUE or FALSE (options A or B).

p	q	$p \wedge r$
T	T	T
T	F	F
F	T	F
F	F	F

Let us construct a truth table for the statement $p \wedge r$. Then we mark the rows in the table in which r is TRUE — rows 1 and 3; given that statement p is FALSE, only row 3 matches the given information. Hence we conclude that the statement $p \wedge r$ is FALSE. This corresponds to option B, which is marked as the correct answer.

Next, we examine whether $p \lor r$ is TRUE or FALSE (options C or D).

It is given that statement r is TRUE and that statement p is FALSE. We construct a truth table for the statement $p \lor r$ and mark the third row, which matches these conditions. We can now conclude that the statement $p \lor r$ is TRUE. This corresponds to option C, which is marked as the correct answer.

p	q	$p \lor r$
T	T	T
T	F	T
F	T	T
F	F	F

Worksheet 3g and Proposed Solutions

Alice was puzzled and distressed by the strange day she had had. This distress led her to have a conversation with herself (pages 16–17).

Alice Wonders who she really is

Alice took up the fan and gloves, and, as the hall was very hot, she kept fanning herself all the time she went on talking! "Dear, dear! How queer everything is to-day! And yesterday things went on just as usual. I wonder if I've been changed during the night?

Let me think: *was* I the same when I got up this morning? I almost think I can remember feeling a little different.

But if I'm not the same, the next question is, who in the world am I?

Ah, *that's* the great puzzle!"

And she began thinking over all the children she knew that were of the same age as herself,

to see if she could have been changed for any of them. "I'm sure I'm not Ada," she said,

"for her hair goes in such long ringlets, and mine doesn't go in ringlets at all; and I'm sure I can't be Mabel, for I know all sorts of things, and she, oh! she knows such a very little!

Besides, *she's* she, and *I'm* I, and — oh dear, how puzzling it all is!

I'll try if I know all the things I used to know. Let me see: four times five is twelve, and four times six is thirteen, and four times seven is — oh dear! I shall never get to twenty at that rate! However, the Multiplication Table doesn't signify: let's try Geography. London is the capital of Paris, and Paris is the capital of Rome, and Rome — no, that's all wrong, I'm certain! I must have been changed for Mabel!"

Each item below contains two statements followed by a question.

Circle the correct answer and explain your reasoning based on the given information.

1. In Wonderland the following statement is TRUE:
    "Ada's hair goes in ringlets and Alice's hair doesn't go in ringlets."

    In Wonderland the following statement is FALSE:
    "Ada's hair goes in ringlets or Alice doesn't know the Multiplication Table."

    Which of the following is correct?

    A. The statement "Alice's hair goes in ringlets" is TRUE.

    B. The statement "Alice's hair doesn't go in ringlets" is FALSE.

    C. **The two given statements contradict one another.**

    D. There is insufficient information given to draw conclusions about Alice's hair.

    Let us designate the statements as follows:

    p: Ada's hair goes in ringlets.

    q: Alice's hair doesn't go in ringlets.

    r: Alice doesn't know the Multiplication Table.

    Given: The compound statement p ∧ q is TRUE. Therefore both p and q are TRUE. But it is also given that p ∨ r is FALSE, and therefore both p and r are FALSE. Thus we obtain that p is both a TRUE statement and a FALSE statement, a condition which cannot exist because of the Law of Excluded Middle (mentioned in Section 2.6). Hence the two given statements contradict one another. This corresponds to option C, which is marked as the correct answer.

2. In Wonderland the following statement is FALSE:
    "Alice knows many things or Mabel knows many things."

    In Wonderland the following statement is TRUE:
    "Alice knows many things or Alice does not know Geography."

Which of the following is correct?

(A.) **The statement "Alice does not know Geography" is TRUE.**

B. The statement "Alice does not know Geography" is FALSE.

C. The two given statements contradict one another.

D. There is insufficient information given to draw conclusions about Alice's knowledge of Geography.

Let us designate the statements as follows:

p: Alice knows many things.

q: Mabel knows many things.

r: Alice does not know Geography.

Given: The compound statement p ∨ q is FALSE. Therefore its two components — the simple statements p, q — are FALSE. It is also given that p ∨ r is TRUE, therefore at least one of its two components — the simple statements p, r — is TRUE. Since p is FALSE, r is necessarily TRUE. This corresponds to option A, which is marked as the correct answer.

3. Both of the following statements are TRUE in Wonderland:

   (i) Alice's hair goes in ringlets, or Mabel knows many things.
   (ii) Alice's hair doesn't go in ringlets, and Alice doesn't know the Multiplication Table.

Conclusions:

A. The statement "Mabel doesn't know many things" is TRUE.

(B.) **The statement "Mabel doesn't know many things" is FALSE.**

C. The two given statements contradict one another.

D. There is insufficient information to draw conclusions on Mabel's knowledge.

Let us designate the statements as follows:

p: *Alice's hair goes in ringlets.*

q: *Mabel knows many things.*

r: *Alice doesn't know the Multiplication Table.*

Given: The compound statement $p \lor q$ is TRUE. That is, at least one of its two components p, q is TRUE. It is also given that the compound statement $\sim p \land r$ is TRUE. That is, its two components $\sim p$, r are TRUE. Since $\sim p$ is a TRUE statement, p is a FALSE statement; thus q is necessarily TRUE. Therefore $\sim q$ is FALSE. This corresponds to option B, which is marked as the correct answer.

3.11. Worksheets 3h and 3i: Summary Exercise for the Logical Connective OR

The summary exercise is composed of two Worksheets: 3h and 3i.

Remarks:

- Worksheet 3h is to be completed as individual practice by students. In Worksheet 3i (which appears in the students' workbook only) students will compare their current answers with the answers they gave at the start of the chapter (Worksheet 3a).
- Upon completion of the summary exercise, it is recommended that a whole class discussion will be held to consider the changes that have taken place in students' understanding and perceptions through the course of this chapter, as well as to identify the particular difficulties encountered with the subject matter.

Worksheet 3h and Proposed Solutions

Answer the questions in this worksheet in their entirety, providing as much detail as possible. If necessary, you may indicate: "I did not understand the question, therefore I have not answered it." Make sure not to go back to Worksheet 1a before you complete your work on this worksheet.

1. You walk into a restaurant and see the following written on the menu: "For the first course you may order soup or salad from the following soup list and salad list..." What do the menu instructions mean? Circle the best answer.

    A. If you order soup you may not order salad.

    B. If you order salad you may not order soup.

    C. You may order both soup and salad.

    D. You may order neither soup nor salad.

    E. There is insufficient information given to determine whether you may order soup and salad.

    F. **More than one of the above answers is correct (indicate which).**

    Your reasoning:

    As indicated in the Logical and Mathematical Background (Section 3.7), the word OR is sometimes used to indicate Exclusive OR. That is, when given a compound statement formed from two simple statements which are connected by the logical connective OR, the context must be examined in order to determine whether the intent is Exclusive OR (recall that the default is Inclusive OR). In the case above, the context indicates that the intent is Exclusive OR. That is, the intent of the restaurant owners is that patrons may order only one of the two. Therefore there is more than one right answer — both A and B are correct. Thus, option F is marked as the correct answer.

2. A mother says to her daughter: "You must eat soup or salad for lunch." The daughter chooses to eat salad. Which of the following statements do you think is correct? Circle the best answer.

    A. The daughter will not eat soup.

B.   The daughter will also eat soup.

C.   **There is insufficient information to determine whether or not the daughter will eat soup.**

Your reasoning:

Linguistically, the case described in this question is very similar to that described in the previous question. What distinguishes them is the context. Unlike the norms in a restaurant, when the mother tells her daughter: "You must eat soup or salad for lunch", it is clear that her intent is "You must eat at least soup or salad." That is, you must eat at least one of the two. If the daughter chooses to eat salad, it may be assumed that her mother will not object to her eating soup as well if she wishes to do so. Hence we do not know if the daughter will eat soup in addition to the salad. This is the common use of OR.

3.   A father says to his son: "Stop making noise or you will be punished." The son stops making noise. Will he be punished? Circle the best answer.

A.   Yes

B.   No

C.   It is reasonable to assume that he will be punished.

D.   **It is reasonable to assume that he will not be punished.**

Your reasoning:

In this case we assume based on context that the father's intent was Exclusive OR. That is, it is reasonable to assume that the father does not mean that if his son stops making noise he will punish him anyway (after all, this is not Wonderland...). This case would apply if the OR in use were the Inclusive OR.

4.   Does the word OR have only a single meaning, or does it mean more than one thing?

If you think OR has only one meaning, write down that meaning in your own words. If you think OR has more than one meaning, write down the different meanings and the distinctions between them.

Your Reasoning:

From the above we see that in spoken language we may attribute two meanings to the word OR.

A. *A conjunction between two options for which **at least** one is TRUE.*

B. *A conjunction between two options for which **exactly** one of them is TRUE.*

By default the first of these two meanings applies.

Sometimes, based on context, we may conclude that the intent is the second meaning.

Chapter 4

De Morgan's Laws

**Neither the Rabbit
nor Alice picks daisies**

4.1. Worksheet 4a: Opening Exercise. Lead-in to De Morgan's Laws and Equivalent Statements

So far we have addressed three logical connectives: NOT, AND, and OR. This chapter links these three connectives without reading on in *Alice's Adventures in Wonderland.*

Instruction of this chapter begins with Worksheet 4a: Opening Exercise (hereforth abb. WS4a) to be completed as individual practice by each student. Upon completion of the task, notify the students that they will return to this worksheet at the end of the chapter (Worksheet 4c Summary Exercise); then they will be able to compare their initial understanding with that which they acquired during the course of the chapter, while reflecting upon the changes in their understanding along with the points where further clarification is still required (Worksheet 4d, which appears only in the Student's Workbook). For this reason it is recommended that discussion of this worksheet will be held off until the end of the chapter. In some cases the teacher may prefer to collect WS4a and give it back only upon completion of WS4c when students are ready to work on WS4d.

Because this lead-in worksheet is repeated again as a summary exercise, the sheet (together with proposed solutions) appears only once — at the end of this chapter.

4.2. Worksheet 4b: Discovering De Morgan's Laws

The objective of Worksheet 4b is to allow students to discover De Morgan's laws on their own using a truth table. At this point students have not yet learned the formal meaning of equivalent statements, but their informal knowledge should suffice as a basis for the topic.

Remarks:

- Worksheet 4b is to be completed in pairs or in small groups to allow for consultation.
- It is suggested that upon completion of the worksheet, students present their answers to the class, in order to initiate a whole class discussion of the solutions.
- At this point it is advised to avoid any judgment of students' answers; points of disagreement and misperceptions, however, should be noted.
- After presentation of the logical and mathematical background by the teacher (Section 4.3 below), students' answers should be revisited and discussed once again.

Worksheet 4b and Proposed Solutions

1. Two statements are considered equivalent if and only if they have identical truth tables. What statement is equivalent to the statement ~(~p)? How can this equivalence be exhibited using a truth table?

The statement ~(~p) is equivalent to the statement p. This can be seen in the table on the right:

p	$\sim q$	$\sim(\sim p)$
T	F	T
F	T	F

The first and last columns are identical, indicating the equivalence of p and ~(~p).

2. Complete the table below (*p* and *q* represent simple statements):

1	2	3	4	5	6	7	8	9	10	11	12
p	$\sim p$	q	$\sim q$	$p \wedge q$	$p \vee q$	$p \wedge (\sim q)$	$(\sim p) \wedge (\sim q)$	$p \vee (\sim q)$	$(\sim p) \vee (\sim q)$	$\sim(p \wedge q)$	$\sim(p \vee q)$
T	F	T	F	T	T	F	F	T	F	F	F
T	F	F	T	F	T	T	F	T	T	T	F
F	T	T	F	F	T	F	F	F	T	T	F
F	T	F	T	F	F	F	T	T	T	T	T

The completed table is as indicated. The Student's worksheet does not include the answers in the 4 bottom rows, of course.

3. Can you identify pairs of equivalent statements in the table?

Columns 8 and 12 are identical, as are columns 10 and 11; therefore the statements that appear at the head of these pairs of columns are equivalent.

4. A. By convention, equivalence is denoted by the symbol ≡.

    Write the equivalent statements that you found in your answer to the previous question using conventional symbols of logic.

    $$(\sim p) \wedge (\sim q) \equiv \sim(p \vee q)$$
    $$(\sim p) \vee (\sim q) \equiv \sim(p \wedge q)$$

    B. What can be conjectured about other relationships between two equivalent statements?

Each one of two equivalent statements follows immediately from the other. Any two equivalent statements say the same thing in two different ways. They may be substituted for one another, because whatever follows from one, follows from the other, and conversely.

5. A. Does your answer to Question 4A remind you of relationships you learned about in set theory between the complement of the union (or intersection) of two sets and the intersection (or union) of their complements?

    Possible solution:
    $$\overline{A \cap B} = \overline{A} \cup \overline{B},$$
    $$\overline{A \cup B} = \overline{A} \cap \overline{B}.$$

    B. If you have learned Venn diagrams before: how can you represent the equivalence you indicated in your answer to part A using Venn diagrams?

    For students who have not yet been exposed to Venn diagrams, a short explanation of them may be needed prior to moving ahead with this question, as per Chapter 5 Section 5.4D.

    Possible solution:

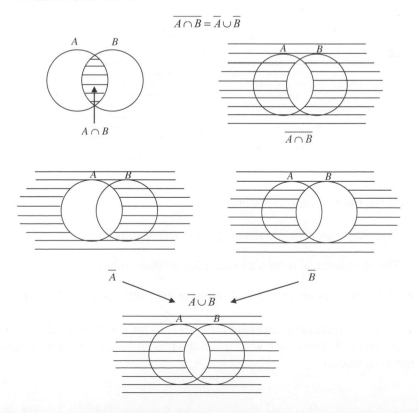

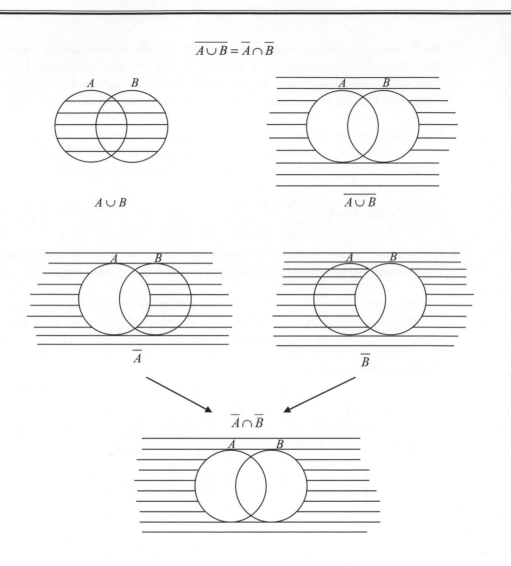

$$\overline{A \cup B} = \overline{A} \cap \overline{B}$$

6.   A. Given the following two statements:

p: The diagonals of a parallelogram are perpendicular.

q: The diagonals of a parallelogram bisect each other.

What is the truth value of $p \wedge q$? What is the truth value of $\sim(p \wedge q)$?

Construct a statement equivalent to $\sim(p \wedge q)$ as you did in your answers to Questions 2, 3, and 4, and write down the two equivalent statements in words (not in symbols).

In reading back your equivalent statements, are you convinced that they do in fact have identical truth values?

The truth value of p is F since the statement is FALSE. This fact suffices for us to state with confidence that the truth value of p ∧ q is also F (a FALSE statement). Thus, its negation ~(p ∧ q) is a TRUE statement (Its truth value is: T).

The statement equivalent to ~(p ∧ q) is (~p) ∨ (~q), or, stated in words:

~(p ∧ q): "It is not true that diagonals in a parallelogram are perpendicular to each other and bisect each other."

(~p) ∨ (~q): "The diagonals in a parallelogram are not perpendicular or they do not bisect each other."

The first verbal statement is quite readable and it is easy to understand that it is a TRUE statement.

The second statement, on the other hand, is harder to understand. It is easier to be convinced of its truth through considerations of logic; that is, from the fact that this is a compound statement that includes the connective OR. The first part is a TRUE statement, and that is enough to state with certainty that its overall truth value is T (It is a TRUE statement).

B. Given two statements:

p: A perfect square number has an even number of divisors.

q: The last digit of a perfect square number can be 2.

What is the truth value of $p \vee q$? What is the truth value of $\sim(p \vee q)$?

Construct a statement equivalent to $\sim(p \vee q)$ as per the instructions given in Questions 2, 3, and 4, and write down the two equivalent statements in words (not in symbols).

In reading back your equivalent statements, are you convinced that they do in fact have identical truth values?

The truth value of p is F since the statement is FALSE. As you know, the number of divisors of a perfect square n = k² is odd (because each divisor smaller than k has a "mate" greater than k, the product of which is n. The divisor k, also called the root, is its own "mate".)

The statement q is also a FALSE statement (since the square of each of the digits 0–9 does not have 2 as its last digit). Thus we can state with confidence that the truth value of $p \vee q$ is also F (a FALSE statement); therefore the truth value of the negation of the statement — $\sim(p \vee q)$ — is T (It is a TRUE statement).

The statement equivalent to $\sim(p \vee q)$ is $(\sim p) \wedge (\sim q)$, or, stated in words: $\sim(p \vee q)$: "It is not true that the number of divisors of a perfect square is even or the last digit of a perfect square could be 2."

$(\sim p) \wedge (\sim q)$: "The number of divisors of a perfect square is odd and the last digit of a perfect square cannot be 2." This is a compound statement that uses the connective AND. Since both components are TRUE, the truth value of the compound statement is T (It is a TRUE statement). In this case the first verbal description is harder to understand than the second.

The answers to this question indicate that verbal formulations using the connective OR are often harder to understand than the equivalent wording that uses the connective AND. De Morgan's laws, which the students have discovered and which will be stated formally later, allow us to move from one formulation to an equivalent one, thereby facilitating better comprehension.

4.3. Logical and Mathematical Background

Students are already familiar with the three logical connectives — NOT, AND, and OR, and the truth table of each. De Morgan's laws create an interesting connection among statements that include them.

Upon completion of Worksheet 4b and after discussion of students' answers, the following concepts and topics are to be presented:

    A. Equivalent statements
    B. De Morgan's laws

A. Equivalent Statements

Two statements p and q are referred to as equivalent if and only if each can be deduced from the other. One method of proving equivalence between two statements is to examine their truth tables. If the tables are identical — the statements are equivalent, and if not — the statements are not equivalent. Two equivalent statements have identical meaning although that meaning may be expressed in two completely different ways, one of which may be easier to understand than the other. From equivalent statements, identical conclusions can be derived. Equivalence is denoted using the $\equiv$ sign, which looks similar to the equal sign; this is not coincidental.

Below is an example of the ideas presented above.

- From the statement $p \wedge q$, it follows that $q \wedge p$. (This can be stated instead as "if $p \wedge q$ then $q \wedge p$".) From the statement $q \wedge p$, it follows that $p \wedge q$. (This can be stated instead as "if $q \wedge p$ then $p \wedge q$".)

  Therefore $p \wedge q \equiv q \vee p$. (This can be stated instead as $p \wedge q$ if and only if $q \wedge p$.)

  Similarly, $p \vee q \equiv q \vee p$.

B. De Morgan's laws

De Morgan's laws are named for the British mathematician and logician August de Morgan (1806–1871), who was the first to prove them. The laws establish the following two equivalences:

    (i) $\sim(p \wedge q) \equiv (\sim p) \vee (\sim q)$
    (ii) $\sim(p \vee q) \equiv (\sim p) \wedge (\sim q)$

This equivalence relation is proven below by using truth tables.

p	$\sim p$	q	$\sim q$	$p \wedge q$	$\sim(p \wedge q)$	$(\sim p) \vee (\sim q)$
T	F	T	F	T	F	F
T	F	F	T	F	T	T
F	T	T	F	F	T	T
F	T	F	T	F	T	T

In the above truth table, the last two columns are identical, proving the first of De Morgan's laws.

In the truth table below, the last two columns are identical, proving the second of De Morgan's laws.

p	$\sim p$	q	$\sim q$	$p \vee q$	$\sim(p \vee q)$	$(\sim p) \wedge (\sim q)$
T	F	T	F	T	F	F
T	F	F	T	T	F	F
F	T	T	F	T	F	F
F	T	F	T	F	T	T

Sometimes it is convenient to understand a statement by examining its equivalent as per De Morgan's laws. An example of this will be shown later in Question 7 of Worksheet 6b.

Venn diagrams[1] aid in the understanding of De Morgan's laws, as indicated in the answer to Question 5 in Worksheet 4b above.

Note: Synonyms in mathematics: Preciseness and lack of ambiguity of each word are among the pillars of mathematical language. There is, however, an exception to this rule (in the sense of "an exception that proves the rule"). By convention, the term "theorem" is used in mathematics for a proven assertion (for example, the

---

[1] As stated above, it is assumed that students are already familiar with the use of Venn diagrams from previous courses in set theory. If they lack this background, a short introduction to the topic is advised.

Pythagorean Theorem). The term "law", however, (e.g. the Commutative Law of Addition), is used to mean the same thing, and similarly the term "rule" (e.g. the Multiplicative Rules). All three of these words — "theorem", "rule" and "law" — are thus mathematical synonyms. This is no doubt an exceptional phenomenon in mathematics. De Morgan's laws are by all means theorems or rules, even though they are called "laws."

4.4. Review of Answers to Worksheet 4b

After students have become familiar with the logical background, it is advised to ask them to go back and review their answers to Worksheet 4b and correct them if necessary.

4.5. Worksheets 4c and 4d: Summary Exercise for De Morgan's Laws

The summary exercise is composed of two Worksheets: 4c and 4d.

Remarks:

• Worksheet 4c is to be completed as individual practice by students. In Worksheet 4d (which appears in the students' workbook only) students will compare their current answers with the answers they gave at the start of the chapter (Worksheet 4a).
• Upon completion of the summary exercise, it is recommended that a whole class discussion be held to consider the changes that have taken place in students' understanding and perceptions through the course of this chapter, as well as to identify the particular difficulties encountered with the subject matter.

Worksheet 4c and Proposed Solutions

Answer the questions in this worksheet in their entirety, providing as much detail as possible. If necessary, you may indicate: "I did not understand the question, therefore I have not answered it." Make sure not to go back to Worksheet 4a before you complete your work on this worksheet.

1.   What is the negation of the following statement:

   "Emma is a dance instructor or a scout leader"?

   A.   Emma is not a dance instructor and is a scout leader.

   B.   Emma is not a dance instructor or is not a scout leader.

   C.   **Emma is not a dance instructor and is not a scout leader.**

   D.   Emma is not a dance instructor or is a scout leader.

   E.   Emma is a dance instructor and is a scout leader.

   F.   Emma is a dance instructor or is not a scout leader.

   G.   Emma is a dance instructor and is not a scout leader.

   Let us designate the statements as follows:

   p: Emma is a dance instructor.

   q: Emma is a scout leader.

   The given statement is $p \vee q$. Its negation is $\sim(p \vee q)$. According to De Morgan's laws, this last statement is equivalent to the statement $(\sim p) \wedge (\sim q)$, expressed verbally as: Emma is not a dance instructor and Emma is not a scout leader. This corresponds to option C, which is marked as the correct answer.

2.   What is the negation of the statement:

   "Ethan is a good athlete and a poor student"?

   A.   Ethan is a poor athlete and a good student.

   B.   Ethan is a poor athlete and a poor student.

C.   Ethan is a good athlete and a good student.

D.   Ethan is a good athlete or a good student.

E.   Ethan is a good athlete or a poor student.

F.   Ethan is a poor athlete or a poor student.

G.   **Ethan is a poor athlete or a good student.**

Let us designate the statements as follows:

p: Ethan is a good athlete.

q: Ethan is a poor student.

The given statement is $p \wedge q$. Its negation is $\sim(p \wedge q)$. According to De Morgan's laws, this last statement is equivalent to the statement $(\sim p) \vee (\sim q)$, expressed verbally as: Ethan is not a good athlete (that is, Ethan is a poor athlete), or Ethan is not a poor student (that is, Ethan is a good student). This corresponds to option G, which is marked as the correct answer.

Chapter 5

The Empty Set

The set Alice and the
Rabbit are elements of,
is not an empty set

Introduction

This chapter focuses on the concept "set," while emphasizing the "empty set," the "finite set," and the "infinite set." Traditionally, these topics appear in a course on set theory, which is often taught before a basic course in logic. Even so, the topics included in this chapter are not a necessary prerequisite for students to be able to progress with this logic course, but their presentation does have (at least) two advantages: these elements provide a good supplement to the logical subject matter presented in this book that mesh well with the flow of the first chapter of *Alice's Adventures in Wonderland*. In addition, they are likely to ease the integration of the topics that will be addressed beginning in Chapter 7 (the quantifiers ALL, THERE EXISTS, and ONLY).

5.1. Worksheet 5a: Opening Exercise. Lead-in to Sets and Types of Sets

Instruction of this chapter begins with Worksheet 5a: Opening Exercise (hereforth abb. WS5a) to be completed as individual practice by each student. Upon completion of the task, notify the students that they will return to this worksheet at the end of the chapter (Worksheet 5d Summary Exercise); then they will be able to compare their initial understanding with that which they acquired during the course of the chapter, while reflecting upon the changes in their understanding along with the points where further clarification is still required (Worksheet 5e, which appears only in the Student's Workbook). For this reason it is recommended that discussion of this worksheet will be held off until the end of the chapter. In some cases the teacher may prefer to collect WS5a and give it back only upon completion of WS5d when students are ready to work on WS5e.

Because this lead-in worksheet is repeated again as a summary exercise, the sheet (together with proposed solutions) appears only once — at the end of this chapter.

5.2. A Look Inside the Text

Read the excerpt that appears in boldface below from pages 3 and 4 of *Alice's Adventures in Wonderland*. These paragraphs will serve as the basis for discussion of the concept "the empty set."

Either the well was very deep, or she fell very slowly, for she had plenty of time as she went down to look about her, and to wonder what was going to happen next. **First, she tried to look down and make out what she was coming to, but it was too dark to see anything; then she looked at the sides of the well and noticed that they were filled with cupboards and book-shelves: here and there she saw maps and pictures hung upon pegs. She took down a jar from one of the shelves as she passed; it was labelled "ORANGE MARMALADE," but to her disappointment it was empty; she did not like to drop the jar for fear of killing somebody underneath, so managed to put it into one of the cupboards as she fell past it.**

"Well!" thought Alice to herself. "After such a fall as this, I shall think nothing of tumbling down stairs! How brave they'll all think me at home! Why, I wouldn't say anything about it, even if I fell off the top of the house!" (This was very likely true.)

5.3. Worksheet 5b: The Empty Set

The central topic of this chapter — the empty set — kicks off with Worksheet 5b, which begins with the excerpt shown above (Section 5.2).

Remarks:

- Worksheet 5b is to be completed in pairs or in small groups to allow for consultation.
- It is suggested that upon completion of this worksheet, students present their answers to the class, in order to initiate a whole class discussion of the solutions.
- At this point it is advised to avoid any judgment of students' answers; points of disagreement and misperceptions, however, should be noted.
- After presentation of the logical and mathematical background by the teacher (Section 5.4 below), students' answers should be revisited and discussed once again.

Worksheet 5b and Proposed Solutions

1. In what way are the jar of marmalade that Alice found and a jar of marmalade found in a store alike? In what way are they different?

   Similarity:

   In both cases there is a jar labeled "Orange Marmalade".

   Difference:

   The jar that Alice found was empty, while marmalade jars sold in stores are full of marmalade.

2. Was the cupboard that Alice fell past empty **before** she fell past it?

   It cannot be determined, since the text states that "[Alice] noticed that they [the sides of the well] were filled with cupboards and book-shelves...". That is, it doesn't state explicitly whether there was anything in the cupboards — not even the specific cupboard into which Alice put the jar. It is stated explicitly that on one of the shelves was a jar, that Alice took and put into "one of the cupboards as she fell past it."

3. Was the cupboard that Alice fell past empty **after** she fell past it?

   It is clear from the excerpt above that the cupboard was not empty after Alice fell past it, since she put the jar labeled "Orange Marmalade" into it.

5.4. Logical and Mathematical Background

Upon completion of Worksheet 5b and after discussion of students' answers, the following concepts and topics are to be presented:

    A. Sets and set notation
    B. Types of sets
    C. Relations between sets
    D. Venn diagrams

A. Sets and set notation

A **set** in mathematics is a collection of elements. Each element belonging to the set appears in it exactly once. In a set there is no significance to the order of appearance of the elements.

In mathematics we are familiar with sets of numbers (the natural numbers, the integers, the rational numbers, the irrational numbers, the real numbers, the complex numbers, etc.), with sets of functions (polynomial functions, rational functions, exponential functions, etc.), with sets of polygons, etc.

By convention, sets are denoted with capital letters (A, B, C,...); elements or their description are listed in curly brackets.

Examples:

$A = \{$The natural numbers divisible by 6$\}$, $C = \{0, 27.15, 3, 36\}$,
$B = \{$All integers that are less than 4$\}$, $Q = \{$The rational numbers$\}$.

By convention, the elements of a set are denoted with lower case letters (a, b, c). An element could belong to a set or not belong to that set, in accordance with the definition of the set. If element k belongs to set S, it is denoted as $k \in S$. If element k does not belongs to set S, it is denoted as $k \notin S$.

Examples:

$36 \in A$; $36 \in C$; $36 \notin B$; $36 \in Q$; $7 \notin A$; $17 \notin C$; $2 \in B$; $-12 \in Q$

B. Types of sets

There exist many types of sets. We will address only a few of them:

A **finite set** is a set that contains a finite number of elements. The number of elements could be very large (for example, the even natural numbers less than ten million), and it could be very small, even containing just a single element (for example, the set of even prime numbers is {2}). A set that contains a single element is called a "singleton".

Note: It is important to distinguish between a specific number (for example, 2), which is not a set, and a set that contains that number only (for example, {2}).

Note: Any set that contains at least one element is non-empty.

An **empty set** is a set that does not contain even a single element. In mathematical terms, it contains zero elements. It is denoted by {} or ϕ.

Examples:

- The set of even primary numbers greater than 10 is an empty set.

- The set of triangles in the Euclidean plane whose sum of angles is greater than 180° is an empty set.

Note: Since there is only one empty set, it is referred to as *the* empty set (and not *an* empty set).

An **infinite** set is a set that contains an infinite number of elements. For example: the set of natural numbers, the set of integers, the set of rational numbers, the set of real numbers.

Infinite sets may differ from one another with regard to their cardinality; however, we will not be addressing enumeration of elements in infinite sets.

Note: In order to avoid paradoxical cases, by mathematical convention an empty set is defined to be a subset of every non-empty set.

C. Relations between sets

Various binary relations exist between pairs of sets. For instance: **identity**, **inclusion**, **intersection** (with **disjoint sets** as a special case), and **union**. Below we define these relations, and provide examples as well as notes and clarifications to elucidate the concepts. Some of the explanations are in the form of theorems, the proofs of which are beyond the scope of this book.

Identity: Two sets are said to be **identical** or **equal** if they contain the same elements. Namely, they are identical if every element in A is an element in B and every element in B is an element in A. This is denoted by: $A = B$.

Examples:

- A is the set of real numbers satisfying the equation $x^2 - 5x + 6 = 0$.

- B is the set of positive prime numbers that are less than 5.

Note: From the definition of set identity it follows that every set is identical to itself.

Inclusion: Set A is said to be **included** in set B if all elements of A are also elements of B. (Inclusion may also be referred to as **containment**.) This is denoted by: $A \subseteq B$. Set A is called a **subset** of set B.

Note: The fact that A is a subset of B does not mean that there must exist an element of B that is not an element of A. If there exists such an element in B (at least one), then A is said to be a **proper** (or **strict**) subset of B, or that it is **strictly included** in B; this is denoted by $A \subset B$.

Note: If $A \subseteq B$ and $B \subseteq A$, then the two sets are identical: $A = B$.

For example:

- The set of natural numbers N is included in the set of integers I, $N \subseteq I$. Moreover, N is strictly included in I, $N \subset I$, because there exist integers that are not natural numbers (for example, -3).

Note: a set A that is included in set B is either strictly included in B or is identical to it. Therefore the expression $A \subseteq B$ is sometimes read as "A is included in or identical to B".

From the definition of inclusion it follows that every set is included in itself.

The empty set is a subset of every set, but not of itself, because a set that includes the empty set is non-empty: $\phi \neq \{\phi\}$. In other words, the empty set is a subset of every non-empty set.

If there are $n > 0$ elements in a set, then the set has 2^n subsets.

The **intersection** of two sets is the set that includes all the elements common to both sets, and only them. The intersection is denoted by $C = A \cap B$.

For example:

- $A = \{\text{The set of numbers divisible by 2}\}$; $B = \{\text{The set of numbers divisible by 3}\}$; $C = A \cap B = \{\text{The set of numbers divisible by 6}\}$.

Note: If $A \subset B$, then $A = A \cap B$.

Note: If two sets have no common elements, they are referred to as **disjoint**. In this case, their intersection is the empty set. This is denoted by $A \cap B = \phi$.

For example:

- $A = \{\text{all even numbers}\}$; $B = \{\text{all odd numbers}\}$; A and B are disjoint sets.

The **union** of two sets is the set that contains all the elements that belong to both sets, and only them. The union is denoted by $C = A \cup B$.

For example:

- $A = \{1, 2, 4, 7, 9\}$; $B = \{1, 4.12, 6, 8, 9\}$; $C = \{1, 2, 4, 4.12, 6, 7, 8, 9\}$.

Note: From the definition of a set, it is clear that in the case where two sets have identical elements, the union contains each element exactly once.

Note: Regarding the number of elements (n) in a set:

If two sets A and B are disjoint, then the following condition holds: $n(A \cup B) = n(A) + n(B)$.

If two sets A and B are not disjoint, then the following condition holds:

$n(A \cup B) = n(A) + n(B) - n(A \cap B)$.

For example:

- $n(A) = 5$, $n(B) = 8$, $n(A \cap B) = 4 \Rightarrow n(A \cup B) = 9$;

- $n(A) = 12$, $n(B) = 27$, $n(A \cup B) = 34 \Rightarrow n(A \cap B) = 5$;

- $n(B) = 7$, $n(A \cap B) = 3$, $n(A \cup B) = 20 \Rightarrow n(A) = 16$;

- $n(A \cap B) = 0$, $n(A \cup B) = 35$, $n(A) = 13 \Rightarrow n(B) = 22$.

D. Venn diagrams

A **Venn diagram** is a useful tool for describing logical relations between sets.

Example:

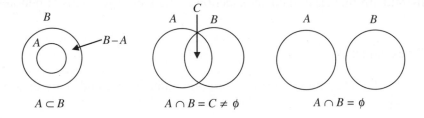

Note: Given $n > 0$ sets, the Venn diagram partitions the plane into 2^n regions, including the plane itself.

Note: $A–B$ denotes the set whose members belong to A and not to B.

Example:

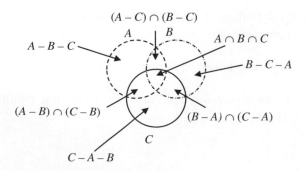

5.5. Review of Answers to Worksheet 5b

After students have become familiar with the logical background, it is advised to ask them to go back and review their answers to Worksheet 5b and correct them if necessary.

5.6. Scientific Discussion: Free Fall and Gravity

In the book *The Annotated Alice* by Martin Gardner, an endnote appears after the words "as she fell past it" (as per the excerpt in Section 5.2 above). In the endnote, the author states the following:

> **Carroll was aware, of course, that in a normal state of free fall Alice could neither drop the jar (it would remain suspended in front of her) nor replace it on a shelf (her speed would be too great).**

To expand the students' knowledge, it is advised to take time out for a discussion of various scientific aspects of free fall.

Free fall is the motion of a body that results from the force of gravity. Gravity, or gravitation, is a natural phenomenon; it is one of the forces in nature. It is the force that attracts bodies to one another with a force proportional to the product of their masses. Gravitational force was discovered by the British physicist and mathematician Sir Isaac Newton (1643–1727). Before Newton's time, the Italian physicist, astronomer and philosopher Galileo Galilei (1564–1642) claimed, as a result of his experiments and observations, that when a body is thrown upward, its velocity decreases at a constant rate, identical to the rate of its increase when the body is falling. The force that acts on that body is what causes the change in its velocity. This is gravitational force.

The acceleration of free fall is not dependent on the mass of the falling body, but rather on the strength of its gravitational field. This acceleration is denoted by the letter g, and its value is approximately 10 meters per second squared (10 m/s^2).

The students may be encouraged to search for additional information on gravitation and free fall, and to address questions such as:

- What is known about the lives of Sir Isaac Newton and Galileo Galilei?
- Is the acceleration of free fall identical in all places on Earth? On mountain tops? Below sea level?
- What additional forces exist in nature?

5.7. Worksheet 5c: Sets and the Number of Elements in a Set

Worksheet 5c addresses sets, relations between sets and the number of elements in a set.

Remarks:

- This worksheet may be completed as individual practice, in pairs, or in small groups, with group discussion encouraged.
- Upon completion of the worksheet, it is recommended that the students present their answers to the class to initiate a whole class discussion of the solutions.

Worksheet 5c and Proposed Solutions

1. In Wonderland there are 30 Pink Rabbits. Alice threw a party and invited all the Pink Rabbits.

    Alice served the Pink Rabbits cheesecake and apple crumb cake.

     Every Pink Rabbit ate at least one piece of cake.

    12 Pink Rabbits ate cheesecake but did not eat apple crumb cake.

    17 Pink Rabbits ate cheesecake.

    A. How many Pink Rabbits ate apple crumb cake?

    Let us define our sets as follows:

    A - The set of Pink Rabbits that ate cheesecake

    B - The set of Pink Rabbits that ate apple crumb cake

    Given: 17 Pink Rabbits ate cheesecake. That is, n(A) = 17.
    Twelve of them did not eat apple crumb cake. That is,
    5 Pink Rabbits ate cheesecake and apple crumb cake.

    Therefore:
    $n(A \cap B) = 5$.

    In order to find $n(B)$, we use the theorem:
    $n(A \cup B) = n(A) + n(B) - n(A \cap B)$.

    We substitute:
    $30 = 17 + n(B) - 5 \Rightarrow n(B) = 18$.

    B. How many Pink Rabbits ate apple crumb cake but did not eat cheesecake?

    We have found that 18 Pink Rabbits ate cheesecake. Since 5 Pink Rabbits ate apple crumb cake and cheesecake, we derive that 13 Pink Rabbits ate apple crumb cake but did not eat cheesecake.

2.   In the school in Wonderland the students may participate in the following extracurricular activities: Carrot cake baking; free fall into the well; orange marmalade making. Each student must choose two of the three. In one of the classes in the school there are 20 students. Ten of them take carrot cake baking, 9 take free fall into the well, and 7 take orange marmalade making.

Four of the students chose both carrot cake baking and free fall into the well. None of the students chose both carrot cake baking and orange marmalade making.

How many students chose both free fall into the well and orange marmalade making but did not choose carrot cake baking?

Let us define our sets as follows:

A - The set of students in carrot cake baking
B - The set of students in free fall into the well
C - The set of students in orange marmalade making

We draw a Venn diagram describing the relations among the three sets, indicating our data accordingly. We also denote by x the number of students that chose free fall into the well as well as orange marmalade making, but did not choose carrot cake baking.

The question may be stated as: What is the value of
$x = n[(B - A) \cap (C - A)]$?

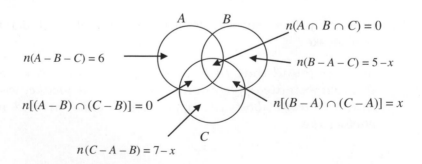

$n(A \cap B \cap C) = 0$

$n(A - B - C) = 6$

$n(B - A - C) = 5 - x$

$n[(A - B) \cap (C - B)] = 0$

$n[(B - A) \cap (C - A)] = x$

$n(C - A - B) = 7 - x$

Reasoning: Since the total number of students in the class is 20, the following condition holds:

$$n(A \cup B \cup C) = 20.$$

Since each student must choose two activities and may not choose three, the following condition holds true:

$$n(A \cap B \cap C) = 0.$$

It is also known that $n(A) = 10$ and that set A has elements in common only with set B; that is:

$$n(A \cap B) = 4; \ n(A \cap C) = 0.$$

From here we obtain that the number of students who participate in carrot cake baking alone is:

$$n(A) - n(A \cap B) - n(A \cap C) = 10 - 4 - 0 = 6.$$

Therefore:

$$n(B - A - C) = 9 - 4 - x = 5 - x,$$

$$n(C - A - B) = 7 - x.$$

Now we solve for x by indicating the total number of students who participate in the activities:

$$6 + 4 + (5 - x) + x + (7 - x) = 20.$$

We solve this equation and find the value of x.

$$\Rightarrow 22 - x = 20 \Rightarrow x = 2.$$

We thus find that two students take free fall into the well as well as orange marmalade making.

Another solution (without using a Venn diagram):

$$n(A) + n(B) + n(C) = 26.$$

That is, there are six "extra" students. Four of them belong to set A and to set B, and were therefore counted twice. Thus two "extra" students remain that were counted twice, and therefore these two students necessarily take free fall into the well as well orange marmalade making.

3.   In Wonderland there are two membership clubs that have established acceptance criteria for those who wish to join.

Acceptance criteria for Club A are:

1.   Being a member of the Wonderland Sports Team

     Or

2.   Having a fondness for orange marmalade and also being a player in the Wonderland Symphony.

Acceptance criteria for Club B are:

1.   Being a player in the Wonderland Symphony

     And

2.   Being a member of the Wonderland Sports Team or having a fondness for orange marmalade.

To which membership club is it more difficult to be accepted?

Let us define our sets as follows:

A - The set of members of the Wonderland Sports Team
B - The set of those who have fondness for orange marmalade.
C - The set of players in the Wonderland Symphony

The acceptance criteria for Club A is to belong to the set: $A \cup (B \cap C)$

The acceptance criteria for Club B is to belong to the set: $C \cap (A \cup B)$

We draw a Venn diagram describing the relations among the three sets, for each of the clubs.

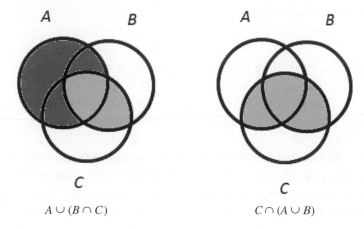

$$A \cup (B \cap C) \qquad\qquad C \cap (A \cup B)$$

It is clear from the Venn diagram that the acceptance criteria for Club B are stricter than those for Club A.

5.8. A Look Inside the Text

Read the two paragraphs that appear in boldface below on pages 4 and 5 of *Alice's Adventures in Wonderland*.

"Well!" thought Alice to herself. "After such a fall as this, I shall think nothing of tumbling down stairs! How brave they'll all think me at home! Why, I wouldn't say anything about it, even if I fell off the top of the house!" (Which was very likely true).

Down, down, down. Would the fall never come to an end? "I wonder how many miles I've fallen by this time?" she said aloud. "I must be getting somewhere near the centre of the earth. Let me see: that would be four thousand miles down. I think — " (for, you see, Alice had learnt several things of this sort in her lessons in the schoolroom, and though this was not a very good opportunity for showing off her knowledge, as there was no one to listen to her, still it was good practice to say it over) " — yes, that's about the right distance — but then I wonder what Latitude or Longitude I've got to?" (Alice had no idea what Latitude was, or Longitude either, but thought they were nice grand words to say.)

Presently she began again. "I wonder if I shall fall right through the earth! How funny it'll seem to come out among the people that walk with their heads downwards! The Antipathies, I think — " (she was rather glad there was no one listening, this time, as it didn't sound at all the right word) " — but I shall have to ask them what the name of the country is, you know. Please, Ma'am, is this New Zealand or Australia?" (and she tried to curtsey as she spoke — fancy curtseying as you're falling through the air! Do you think you could manage it?) "And what an ignorant little girl she'll think me! No, it'll never do to ask: perhaps I shall see it written up somewhere."

5.9. Scientific Discussion: Earth

To expand the students' knowledge, it is advised to take time out for a discussion of various scientific and geographical aspects of earth, such as longitude and latitude, the radius of the earth, etc.

It is advised to begin by recalling the definition of a sphere:

A sphere is the geometric locus of all points in the three-dimensional Euclidean space that are located at a distance R ("the radius") from a given point O ("the center"). A line segment connecting two points on the sphere that passes through the center is called a diameter. Pair of diametrically opposed points on the sphere are called — antipodes.

Below is a short informational passage about the earth (the numbers that appear in the explanation are rounded to simplify understanding).[1]

> **Earth** is a planet in the solar system. It is the third planet from the Sun, and the fifth largest in the system. Earth is the only planet known to have life on it. Scientific evidence indicates that the earth was formed around 4.5 billion years ago, and that shortly thereafter it acquired the single natural satellite that orbits it — the moon. It is hypothesized that the earth's central core reaches temperatures that exceed 10,000 degrees Fahrenheit or 6,000 degrees Celsius. The heat is a result of gravitational force (see also Section 5.6 above), which exerts tremendous pressure; this is then converted into heat energy.
>
> The earth's shape is a spheroid: not quite a sphere because it is slightly squashed on the top and bottom. That is, the diameter that connects the two poles (North and South) is shorter (by some 25 miles) than its diameter at the equator. The diameter at the equator is nearly 8,000 miles, and its equatorial circumference is nearly 25,000 miles. Its circumference around the north–south longitudinal lines is some 45 miles shorter.
>
> The **equator** divides Earth into two halves: the Northern Hemisphere and the Southern Hemisphere. By definition the equator is the 0° latitude line.
>
> A **latitude line**, is an imaginary circle on the surface of the earth, parallel to the equator, which is the imaginary line that orbits the earth halfway between the North and South Poles. The equator is, in effect, the longest latitude line (some 25,000 miles, as stated above). Latitude lines connect all the points that have the same geographical latitude. Latitude is measured in degrees, beginning with 0° at the equator, and reaching 90° at the poles.

---

[1] The data is based on Wikipedia entries for "earth", "longitude" and "latitude."

Latitude coordinates are provided together with an indication of whether they are north or south of the equator.

A **longitudinal line**, or **meridian**, is an imaginary circle on the surface of the earth connecting the North and South Poles, with a length equal to half the circumference of the earth. A meridian connects all the points that have the same geographical longitude. Each meridian crosses the equator, and because the equator is a circle, it may be divided into 360°, thus defining 360 meridians. The zero degree longitude has been arbitrarily selected; it is known as the Prime Meridian and passes through the Royal Observatory in Greenwich, London. Every meridian indicates how far a particular point on earth that lies on it is east or west of the Prime Meridian. The area west of Greenwich is called the Western Hemisphere, while that east of Greenwich is the Eastern Hemisphere.

Any point on earth may be identified by indicating its geographical longitude and latitude pair. Together these define the precise location of the point.

The students may be encouraged to search for additional information and to address questions such as:

- What is the solar system?
- What are the dimensions of the earth?
- How can one navigate using latitude and longitude?
- How do the latitude lines divide the earth into different regional climates?
- What is the relationship between longitude and time zones?
- What do you think would happen if you could fall into the earth?
- Could it be that Alice's fall will never end? Why? (What things do end? What things never end?)

5.10. Discussion of Educational Issues — Showing Off Knowledge and the Search for Information

It is advised to take time out to take a closer look at this excerpt and hold a discussion with the students on various aspects of education that are not necessarily mathematical or scientific. The excerpt quoted above is an opportunity to discuss the need to search for information and the penchant to show off knowledge. The table shown below presents quotes from the excerpt above and the educational issues that arise from them:

Quote	Educational issue arising from it
"[...] and though this was not a very good opportunity for showing off her knowledge, as there was no one to listen to her, still it was good practice to say it over." "(she was rather glad there was no one listening, this time, as it didn't sound at all the right word)"	What is the purpose of showing off knowledge? When does the need arise for showing off knowledge? Should children be educated to show off their knowledge in front of others only when they are certain that their statement is correct, or should they be given the sense that making a mistake is alright as well? What about teaching children to be humble?
"(Alice had no idea what Latitude was, or Longitude either, but thought they were nice grand words to say.)."	Why do children feel the need to say "nice grand words" even when they don't understand their meaning? Should children be educated to avoid this? How can they be encouraged to ask about the meaning of every word they don't understand?
"And what an ignorant little girl she'll think me!"	Why do children sometimes avoid asking questions? Is it important to build children's confidence to ask questions even if they seem like "stupid" questions? What is likely to cause children to feel uneasy and to avoid asking those types of questions?

"No, it'll never do to ask: perhaps I shall see it written up somewhere."	Is it important to educate children to search for information on their own? What could motivate children to discover information on their own? What can arouse in them the will to do so? What are our objectives as teachers, when we send children to search for information on their own? What could hamper the achievement of these goals?

5.11. Worksheets 5d and 5e: Summary Exercise for Sets and Types of Sets

The summary exercise is composed of two Worksheets: 5d and 5e.

Remarks:

- Worksheet 5d is to be performed as individual practice by students. In Worksheet 5e (which appears in the students' workbook only) students will compare their current answers with the answers they gave to the same questions at the start of the chapter (Worksheet 5a).
- Upon completion of the summary exercise it is recommended that a whole class discussion be held in order to consider the changes that have taken place in students' understanding and perceptions through the course of this chapter, as well as to identify the particular difficulties encountered with the subject matter.

Worksheet 5d and Answer Key

Answer the questions in this worksheet in their entirety, providing as much detail as possible. If necessary, you may indicate: "I did not understand the question, therefore I have not answered it." Make sure not to go back to Worksheet 4a before you complete your work on this worksheet.

1. What do you think of when you hear the word "set"? That is, what association does the word "set" bring up to you? Provide at least five examples.

   A set of flatware, a set of carving knives, a chess set, a set in tennis, a sweater set.

2. Is there something common to all the answers you gave for Question 1? If so — what? If not — indicate this.

   The common thing is that the members of each set have some common property that enables them to function or to be used together to achieve a single objective (for example, members of the set of flatware are designed to help people to eat). Note: In spoken language the word "set" is used to group items that are meant to function or to be used together for a particular objective. The function described above is an example of this. In mathematics, too, membership of elements in a particular set is usually determined by a common property, but this is not always the case; membership in a mathematical set is not necessarily based on a common property.

3. Try to define the concept "set".
   Begin your answer with the words: "A set is...".

   A set is a collection of elements (students are likely to indicate that the elements have a "common property", as expected from their real-world experience).

4. Do you think there exists a set that can be called an "empty set"? If so — explain your answer and provide examples; if not — why?

A set that has no elements is an "empty set". Examples of this are the set of four-sided triangles; the set of eligible voters under the age of 15; the set of numbers whose absolute value is negative.

5.   Do you think there is a set that can be called an "infinite set"?
      If so — explain your answer and provide examples; if not — why?

An infinite set is a set with an infinite number of members. Examples of this are the set of natural numbers; the set of integers; the set of rational numbers; the set of irrational numbers; the set of real numbers; even the set of points in a finite line segment.

Chapter 6

Relations and Their Properties

**Alice loves the
White Rabbit.
Does this mean that
the White Rabbit
loves Alice?**

Introduction

As in the previous chapter, the central topic of this chapter — relations and their properties — is not a necessary prerequisite for students to progress with the basic logic course, and it is likely that some of the students are already familiar with the topic from previous mathematics courses. The presentation of this topic, however, does have (at least) two marked advantages: First, it supplements the flow of the text in the first chapter of *Alice's Adventures in Wonderland*, and second, it is likely to ease the integration of the topics that will be addressed beginning in Chapter 7.

6.1. Worksheet 6a: Opening Exercise. Lead-in to Relations and Their Properties

Instruction of this chapter begins with Worksheet 6a: Opening Exercise (hereforth abb. WS6a) to be completed as individual practice by each student. Upon completion of the task, notify the students that they will return to this worksheet at the end of the chapter (Worksheet 6d Summary Exercise); then they will be able to compare their initial understanding with that which they acquired during the course of the chapter while reflecting upon the changes in their understanding along with the points where further clarification is still required (Worksheet 6e which appears only in the Student's Workbook). For this reason it is recommended that discussion of this worksheet will be held off until the end of the chapter. In some cases the teacher may prefer to collect WS6a and give it back only upon completion of WS6d when students are ready to work on WS6e.

Because this lead-in worksheet is repeated again as a summary exercise, the sheet (together with proposed solutions) appears only once — at the end of this chapter.

6.2. A Look Inside the Text

Read the excerpt that appears in boldface below from pages 5 and 6 of *Alice's Adventures in Wonderland*. This excerpt will serve as the basis for discussion for the topic "Relations and their Properties."

Down, down, down. There was nothing else to do, so Alice soon began talking again. "Dinah'll miss me very much to-night, I should think!"

(Dinah was the cat.) "I hope they'll remember her saucer of milk at tea-time. Dinah, my dear, I wish you were down here with me! There are no mice in the air, I'm afraid, but you might catch a bat, and that's very like a mouse, you know. But do cats eat bats, I wonder?" And here Alice began to get rather sleepy, and went on saying to herself, in a dreamy sort of way, "Do cats eat bats? Do cats eat bats?" and sometimes, "Do bats eat cats?" for, you see, as she couldn't answer either question, it didn't much matter which way she put it. She felt that she was dozing off, and had just begun to dream that she was walking hand in hand with Dinah, and saying to her very earnestly, "Now, Dinah, tell me the truth: did you ever eat a bat?" when suddenly, thump! thump! down she came upon a heap of sticks and dry leaves, and the fall was over.

6.3. Worksheet 6b: Relations and Their Properties

The central topic of this chapter — relations and their properties — kicks off with Worksheet 6b, which begins with the excerpt shown above (Section 6.2).

Remarks:

- Worksheet 6b is to be completed in pairs or in small groups to allow for consultation.
- It is suggested that upon completion of this worksheet, students present their answers to the class, in order to initiate a whole class discussion of the solutions.
- At this point it is advised to avoid any judgment of students' answers; points of disagreement and misperceptions, however, should be noted.
- After presentation of the logical and mathematical background by the teacher (Section 6.4 below), students' answers should be revisited and discussed once again.

Worksheet 6b and Proposed Solutions

1. If it is known that bats are very similar to mice, does it follow that mice are very similar to bats? Why?

    The answer to this question is "Yes". Similarity is a reciprocal relationship. (Note: As will be shown later, such a relation is said to be symmetric.)

2. If it is known that cats eat bats, does it follow from this that bats eat cats? Why?

    The answer to this question is "No". In the animal kingdom, "eats", for the most part, is not a reciprocal relationship but rather a unidirectional one. (Note: as will be shown later, such a relation is said to be non-symmetric.)

3. What is the difference between the relation described in Question 1 and the relation described in Question 2?

    The relation described in Question 1 is reciprocal — if A is similar to B, then B is necessarily similar to A. The relation described in Question 2 is not necessarily reciprocal — A eats B does not imply that B also eats A.

4. It is known that Alice likes Dinah the cat. It is known that Dinah the cat likes the White Rabbit. Does it follow from this that Alice likes the White Rabbit? Why?

    The answer to this question is "No". "Like" is a relation that is not transferred from one party to a third party via an agent. (Note: As will be shown later, the relation in this case is said to be non-transitive.)

5. It is known that Alice is taller than Dinah the cat. It is known that Dinah the cat is taller than the bat. Does it follow from this that Alice is taller than the bat? Why?

    The answer to this question is "Yes". "Tallness" (that is, "height") is measured in numbers. As such, any two heights may be compared and ordered by their magnitude. The order relation between the numbers is transferred from one party to a third party via an agent. (Note: As will be shown later, the relation in this case is said to be transitive.)

6. What is the difference between the relation described in Question 4 and the relation described in Question 5?

    The relation that appears in Question 5 is transferred to a third party via an agent while the relation in Question 4 is not.

7. Lewis Carroll states in the book: "for, you see, as she couldn't answer either question, it didn't much matter which way she put it."

    A. In the phrase "she couldn't answer either question" in the sentence above, what are the two components referred to by the phrase "either question"?

    The components are "do cats eat bats?" and "do bats eat cats?".

    B. Construct the phrase "for ... either question" in statement notation using "or", construct the equivalent statement using De Morgan's Laws, and finally, express the equivalent statement in words. Is this equivalent sentence identical to the original sentence in the context of the story? Why?

    Let us define the following statements:

    p: Cats eat bats.

    q: Bats eat cats.

    The sentence in statement notation: $\sim p \wedge \sim q$.

    The equivalent statement as per De Morgan's Laws is $\sim(p \vee q)$, since $\sim p \wedge \sim q \equiv \sim(p \vee q)$.

    The equivalent sentence states: "not both statements." That is, it negates the statement "Cats eat bats AND bats eat cats." This sentence is identical to the original one, even in the context of the story.

    C. Why did it not matter to Alice which way she put her words? Does order never matter? Explain your answer.

There are cases where the order makes no difference. For example: "A is similar to B" is identical in meaning to "B is similar to A" due to the special property of "similarity." But "A eats B" is not identical in meaning to "B eats A", because "eating" is not a reciprocal operation. (Note: As will be explained later, the first of these relations is said to be symmetric. We may deduce from this that in cases of symmetric relations, the order does not matter. If the relation is not symmetric, changing order changes the meaning.)

6.4. Logical and Mathematical Background

Upon completion of Worksheet 6b, and after discussion of the students' answers, the following concepts and topics are to be presented:

    A. The concept "relation"
    B. Defining a relation
    C. Properties of relations — reflexive, symmetric, transitive

A. The concept "relation"

A **relation** is a connection of some sort that may or may not exist between two objects (not necessarily mathematical), or between an object and itself.

Note: To be precise, what has been mentioned here is a binary relation; that is, a relation between **two** objects (identical or different). We will not be addressing relations among more than two objects.

We are familiar with many relations in mathematics, such as greater than, equal to, successor of, etc. In everyday life we are familiar with relations such as: father of, friend of, similar to, etc.

There is a distinction between relations and properties — relations are always between two objects, while properties are associated with a single object.

For example:

- The properties "composite" and "prime" do not express relations between two objects; rather, they are properties of objects themselves.

B. Defining a relation

Every relation is defined on members of a particular set. For example, the relation "multiple of" is defined on the set of natural numbers and not on a set of letters or a set of people. The relation "grandmother of", on the other hand, is defined on people and not on numbers.

Each relation requires a definition of its own, while considering the set on whose members it is defined.

For example:

- The relation "greater than" on the set of natural numbers is defined as follows: a natural number n is greater than a natural number m if there exists a natural number d satisfying $n = m+d$, and vice versa (namely, given that n and m are natural numbers, if there exists a natural number d satisfying $n = m+d$, then n is greater than m.)[1] On the set of humans, the definition is: A is greater than B if (s)he is of higher stature (this definition is certainly not unequivocal as the mathematical definition is). Mathematicians say that a relation is a collection of pairs of elements that may be created from the elements of a particular set.

Note: A relation could be defined on a particular set, and still not be defined on a set that includes it or that is similar to it in some way.

For example:

- The relation "greater than" is not defined on the set of positive integers modulo 12. This is easy to see if you think of the set of hours on the clock $\{1, 2, 3, 4, 5, 6, 7, 8, 9, 10, 11, 12\}$, and of "greater than" as "later than". It is meaningless to say that 7:00 is later than 3:00, because we can, by the same token, say that 3:00 is later than 7:00. The absence of a unique meaning invalidates the definition of the relation "greater than" on this set.

- The relation "greater than" is defined for the set of integers, but it is not defined on the broader set of complex numbers, which includes all integers.

C. Properties of relations

In a set in which the relation R is defined, the relation R could be reflexive and/or symmetric and/or transitive. We now define these three properties.

---

[1] Mathematicians use the abbreviated style "if and only if" when two statements of the form "if a, then b" as well as "if b, then a" are both true. In such cases they say "a if and only if b". See Chapter 8 for more about "if… then…" statements. See also Chapter 4 Section 4.3.

C.1. Reflexive relation

Definition: A relation R defined on a set S is **reflexive** if and only if the relation R exists between each element a of the set and itself. More formally: for all $a \in S$, aRa.

Examples of reflexive relations:

- The relation: Equality
  The set: The natural numbers
  Every natural number is equal to itself.

- The relation: Congruence
  The set: Triangles in the plane
  Every triangle is congruent to itself.

- The relation: Sibling of (i.e., having two common parents)
  The set: Humans
  Each human is his/her own sibling.

Examples of non-reflexive relations:

- The relation: Greater than
  The set: The rational numbers
  No rational number is greater than itself.

- The relation: Father of
  The set: Humans
  No human male is his own father.

C.2. Symmetric relation

Definition: A relation R, defined on a set S, is **symmetric** if and only if for every two elements a, b in the set, if a is related to b, it follows that b is related to a. It is common to use a double arrow for "if and only if" and state it more formally this way:

for all $a, b \in S$, $aRb \Leftrightarrow bRa$.

Examples of symmetric relations:

- The relation: Equality
  The set: The natural numbers
  In the set of natural numbers, $n = m \Leftrightarrow m = n$ for every two natural numbers m, n.

- The relation: Congruence
  The set: Triangles in the plane
  In the set of triangles in the plane, $\triangle ABC \cong \triangle DEF \Leftrightarrow \triangle DEF \cong \triangle ABC$ where $\triangle ABC$ and $\triangle DEF$ are two triangles in the plane.

- The relation: Sibling of (i.e., having two common parents)
  The set: Humans
  a is the sibling of b if and only if b is the sibling of a, for any two people a and b.

These relations are both symmetric and reflexive. But there also exist symmetric relations that are **not** reflexive.

For example:

- The relation: Perpendicular to
  The set: Lines in the plane
  $l \perp k \Leftrightarrow k \perp l$, for all k, l that are lines in the plane, but a line is not perpendicular to itself.

- The relation: Married to
  The set: Humans
  a is married to b if and only if b is married to a, for every two people a and b, but no person is married to himself, by nature of the definition of marriage.

  Surely not every relation is symmetric.

  Examples of asymmetric relations:

- The relation: Greater than
  The set: The rational numbers
  If a is greater than b, then b is not greater than a, for every a, b that are rational numbers. In this case it is even true that b is less than a. This relation is said to be not only asymmetric, but even antisymmetric.

- Relation: Child of
  The set: Humans
  If a is the child of b, then b is not the child of a, where a, b are human beings. Moreover, it can also be said that b is the parent of a.

- The relation: Square of
  The set: The rational numbers
  If r is the square of q, namely $q^2 = r$ where r, q are rational numbers, it is not necessarily the case that q is the square of r, (it may be, in the case that $q = 0$ or $q = 1$, but it also may not be — in every other case). Therefore it is certain that this relation is asymmetric. Nevertheless, we cannot state with certainty that $q = \sqrt[2]{r}$, because q could be negative, therefore we cannot say that it is antisymmetric.

- The relation: Love
  The set: Humans
  If a loves b, where a, b are humans, it does not follow that b necessarily loves a, therefore the relation 'love' is not symmetric (but it cannot be deduced, of course, that b hates a, therefore this relation is not antisymmetric).

C.3. The transitive relation

Definition: A relation R defined on a set S is **transitive** if and only if for every three elements a, b, and c in the set S, if a is related to b, and b is related to c, it follows that the relation also holds between a and c. More formally: for all $a, b, c \in S$, $aRb \wedge bRc \Rightarrow aRc$.

Examples of transitive relations:

- The relation: Equality
  The set: The natural numbers
  $[(l = m) \wedge (m = n)] \Rightarrow l = n$, where l, m, n are natural numbers.

  Note: As seen above, in addition to being a transitive relation, equality as defined on the set of natural numbers is also symmetric and reflexive. A relation that has all three of these properties is called an **equivalence relation**.

- The relation: Greater than
  The set: The rational numbers
  $[(l > m) \wedge (m > n)] \Rightarrow l > n$, where l, m, n are rational numbers.

  Note: As seen above, the relation "greater than" as defined on the set of rational numbers is neither symmetric nor reflexive. A transitive relation that is neither reflexive nor symmetric is called an **order relation**.

- The relation: Sibling of (i.e., having two common parents)
  The set: Humans
  If a is the sibling of b, and b is the sibling of c, then a is the sibling of c (for any humans a, b, and c).

  Note: This relation, too, is an equivalence relation (why?)

Examples of non-transitive relations:

- The relations paternity and "cousin of" in the set of humans, the relation "mathematical square" in the set of real numbers.

  Note: It is worthwhile noting that the definition of a particular relation on a given set may depend on mathematical operations defined on that set.

For example:

The definition of the relation "greater than" on the set of integers is: for all integers a, b, $a > b$ if and only if there exists an integer $c > 0$ such that $a = b + c$. Obviously, a change to the definition of this operation affects the nature of the relation. Thus, for example, in the set of integers modulo 5 $\{0, 1, 2, 3, 4\}$, the values in the addition table differ from those in the addition table for the set of integers. Hence, a relation similarly defined on the set of integers modulo 5 cannot be called "greater than" in the same sense of the term, since in this set $2 > 3$ while at the same time $3 > 2$. Despite this, it is a well-defined relation, but it has "strange" properties if we think of it as "greater than" in the commonly used sense. For example, 0 is greater than 4, but 0 is not greater than itself, because there does not exist a number in the set $c > 0$, such that $0 + c = 0$. Therefore it is best to think of it as the relation R and to denote it by $2R3$ and $3R2$ rather than by the 'greater than' symbol.

6.5. Review of Answers to Worksheet 6b

After students have become familiar with the logical background, it is advised to ask them to go back and review their answers to Worksheet 6b and correct them if necessary.

It is advised to take time out for a discussion on question 7C of the worksheet — in what cases does order matter, and in what cases does it not? For example, when we add or multiply two numbers, we tend to say "order doesn't matter," since these operations are commutative. But what happens when we subtract or divide two numbers? In this case the order **is** important, since these operations are **not** commutative.

Similarly, when the relation is symmetric, it doesn't matter whether we say "Jill is Kimberly's sister" or "Kimberly is Jill's sister."

6.6. Worksheet 6c: More on Relations and Their Properties

Worksheet 6c is designed to allow students to deepen their understanding of the distinctions between the various properties of relations.

This worksheet has two sections. In the first section, students are presented with sets and relations, and are asked to determine which properties apply to each. In the second section, students are presented with properties and are asked to find relations and sets for which these properties apply.

Remarks:

- This worksheet may be completed as individual practice, in pairs or in small groups, with group discussion encouraged.
- Upon completion of the worksheet, it is recommended that the students present their answers to the class to initiate a whole class discussion of the solutions.

Worksheet 6c and Proposed Solutions

1. Alongside each set shown in the table is a relation. Check which properties hold for that relation as defined on the set and which do not.

The set	The relation	The properties		
		Reflexive	**Symmetric**	**Transitive**
Humans	Brother-in-law of/sister-in-law of	– *No person is his own brother-in-law.*	+ *If k is the brother-in-law/sister-in-law of m, then m is the brother-in-law/sister-in-law of k.*	– *If k is the brother-in-law/sister-in-law of m, and m is the brother-in-law/sister-in-law of n, it does not necessarily follow that k and n are brothers-in-law/sisters-in-law.*
The natural numbers	A divisor of	+ *Every natural numbers is a divisor of itself.*	– *If k is a divisor of m, then m is not necessarily a divisor of k (this is an asymmetric relation).*	+ *If k is a divisor of m (that is, m is a multiple of k), and m is a divisor of n (that is, n is a multiple of m), then it follows that k is a divisor of n (that is, n is a multiple of k).*

(Continued)

(*Continued*)

The set	The relation	The properties		
		Reflexive	Symmetric	Transitive
Countries	Latitude is north of	– *No country lies on a line of latitude north of its own.*	– *If the latitude of country k is north of the latitude of country m, then country m's latitude is not north of the latitude of country k, but rather south of it. (This relation is antisymmetric.)*	+ *If the latitude of country k is north of the latitude of country m, and the latitude of country m is north of the latitude of country n, then the latitude of country k is north of the latitude of country n.*

2. Add examples of your own to the sets and relations as per the properties listed in the table, and explain.

The set	The relation	The properties		
		Reflexive	Symmetric	Transitive
The natural numbers greater than 1	*Having a common factor different from 1*	+	+	–

Reasoning:

The relation is reflexive, because each natural number > 1 has a common factor different from 1 with itself, that being the number itself.

The relation is symmetric, since for each pair of natural numbers k and m greater than 1 of which n is a common factor, then n is also a common factor of m and k.

The relation is non-transitive, since for every three natural numbers k, m and n greater than 1, if k and m have a common factor, and m and n have a common factor, it does not necessarily follow that k and n have a common factor. For example, 4 and 6 share the common factor 2; 6 and 9 share the common factor 3, but 4 and 9 have no common factor other than 1.

The set	The relation	Reflexive	Symmetric	Transitive
Humans	Learning from the experience of	+	−	−

Reasoning:

The relation is reflexive, since every person learns from his own experience.

The relation is not symmetric, since if person k learns from the experience of person m, it does not necessarily follow that person m learns from the experience of person k.

The relation is not transitive, since if person k learns from the experience of person m, and person m learns from the experience of person n, it does not necessarily follow that person k learns from the experience of person n.

The set	The relation	Reflexive	Symmetric	Transitive
Triangles	Congruence	+	+	+

Reasoning:

The relation is reflexive since each triangle is congruent to itself.
The relation is symmetric, since if triangle ABC is congruent to triangle DEF, then triangle DEF must be congruent to triangle ABC.

The relation is transitive, since if triangle ABC is congruent to triangle DEF, and triangle DEF is congruent to triangle GHJ then triangle ABC must be congruent to triangle GHJ.

The set	The relation	Reflexive	Symmetric	Transitive
The real numbers	Square of	–	–	–

Reasoning:

The relation is not reflexive, since not every real number is its own square. For example, 7.15 is not a square of itself. (There are many additional examples; note, however, that there are cases that cannot serve as examples, e.g. 1 is a square of itself, as is zero.)

The relation is not symmetric, since if a real number k is the square of a real number m, it does not necessarily follow that m is the square of k. For example, 2.25 is the square of 1.5, but 1.5 is not the square of 2.25.

The relation is not transitive, since if a real number k is the square of a real number m, and the real number m is the square of a real number n, it does not necessarily follow that the real number k is the square of the real number n.

The set	The relation	Reflexive	Symmetric	Transitive
Does not exist	No relation of this sort exists	–	+	+

Reasoning:

Such a relation cannot exist. The reason is that any relation that is symmetric and transitive is necessarily reflexive as well. This can be explained as follows:

Because it is symmetric, it follows that for all $a, b \in S$, $aRb \Rightarrow bRa$. Because it is transitive, it follows that for all $a, b \in S$, $aRb \wedge bRa \Rightarrow aRa$. But this means that it must be reflexive as well. Hence there is no set on which such a relation can be defined.

6.7. Worksheets 6d and 6e: Summary Exercise for Relations and Their Properties

The summary exercise is composed of two Worksheets: 6d and 6e.

Remarks:

- Worksheet 6d is to be performed as individual practice by the students. In Worksheet 6e (which appears in the students' workbook only) students will compare their current answers with the answers they gave to the same questions at the start of the chapter (Worksheet 6a).
- Upon completion of the summary exercise, it is recommended that a whole class discussion be held in order to consider the changes that have taken place in students' understanding and perceptions through the course of this chapter, as well as to identify the particular difficulties encountered with the subject matter.

Worksheet 6d and Proposed Solutions

Answer the questions in this worksheet in their entirety, providing as much detail as possible. If necessary, you may indicate: "I did not understand the question, therefore I have not answered it." Make sure not to go back to Worksheet 4a before you complete your work on this worksheet.

1. You run into Jason and Andrew sitting together in the cafeteria. List at least six possible relations between Jason and Andrew, for example: family, friendship, competition, height, weight. List as many statements as you can about the relationship between Jason and Andrew in the format: "Jason is ... of Andrew", "Andrew is more ... than Jason", "Jason has less ... than Andrew".

    1. Jason is Andrew's brother.
    2. Jason eats in the cafeteria less often than Andrew.
    3. Andrew is taller than Jason.
    4. Jason is Andrew's instructor.
    5. Jason earns the same salary as Andrew.
    6. Andrew is Jason's grandfather.

2. It is known that Jason has a certain relationship with Andrew (one of those that you listed in your answer to Question 1). Is Andrew necessarily related to Jason in the same way? Explain your answer for each statement you gave for the previous question.

    Not necessarily. If Jason is Andrew's brother (1) then it follows that Andrew is Jason's brother.

    If Jason eats less than Andrew (2), clearly Andrew does not eat less than Jason.

    If Andrew is taller than Jason (3), then Jason is necessarily not taller than Andrew.

    If Jason is Andrew's instructor (4), it is possible that in another school or in another course Andrew is Jason's instructor, but not in the same course.

    If Jason earns the same salary as Andrew (5), it is clear that Andrew earns the same salary as Jason.

    If Andrew is Jason's grandfather (6), it is clear that Jason is not Andrew's grandfather.

3. It is known that Jason is related to Andrew in some way, and that Andrew is related to you in the same way. Does it follow that Jason is related to you in the same way? Explain your answer.

    No. Let us assume that Jason is Andrew's cousin on Andrew's father's side, and Andrew is your cousin on Andrew's mother's side. Thus Jason and you are not necessarily in the relation "cousin of."

4. In the list you prepared in response to Question 1, find a relation that exists between Jason and Andrew, such that if you assume that it also applies between Andrew and you, you may also state with confidence that the relation exists between Jason and you. If you do not have such a relation in your list, add one.

    Relations 1, 2 and 3 are such relations.

    1. *If Jason is Andrew's brother, and Andrew is my brother, then Jason is my brother (full brothers, not half-brothers).*
    2. *If Jason eats in the cafeteria less frequently than Andrew, and Andrew eats in the cafeteria less frequently than I do, then Jason eats in the cafeteria less frequently than I do.*
    3. *If I am taller than Andrew and Andrew is taller than Jason, then I am taller than Jason.*

5. In the list you prepared in response to Question 1, find a relation that exists between Andrew and Jason, such that if you assume that it also applies between Jason and you, you may also state with confidence that the relation does **not** exist between Andrew and you. If you do not have such a relation in your list, add one.

    Relation 6 is such a relation. If Andrew is Jason's grandfather, and Jason is my grandfather, it could not be that Andrew is my grandfather (except for the unusual case where my father is married to my grandfather Jason's daughter...).

Chapter 7

The Quantifiers
ALL, THERE EXISTS, and ONLY

There EXISTS a little
door, that ONLY behind it
could Alice find
ALL she was looking for

7.1. Worksheet 7a: Opening Exercise. Lead-in to the Quantifiers ALL, THERE EXISTS, AND ONLY

Instruction of this chapter begins with Worksheet 7a: Opening Exercise (hereforth abb. WS7a), to be completed as individual practice by each student. Upon completion of the task, notify students that they will return to this worksheet at the end of the chapter (Worksheet 7f Summary Exercise); then they will be able to compare their initial understanding with that which they acquired during the course of the chapter while reflecting upon the changes in their understanding along with the points where further clarification is still required (Worksheet 7g which appears only in the Student's Workbook). For this reason it is recommended that discussion of this worksheet will be held off until the end of the chapter. In some cases, the teacher may prefer to collect WS7a and give it back only upon completion of WS7f when students are ready to work on WS7g.

Because this lead-in worksheet is repeated again as a summary exercise, the sheet (together with proposed solutions) appears only once — at the end of this chapter.

7.2. A Look Inside the Text

Read the three paragraphs from pages 6 and 7 of *Alice's Adventures in Wonderland* that appear in boldface below. They will serve as the basis for discussion of the quantifiers ALL and THERE EXISTS.

Alice was not a bit hurt, and she jumped up on to her feet in a moment: she looked up, but it was all dark overhead; before her was another long passage, and the White Rabbit was still in sight, hurrying down it. There was not a moment to be lost: away went Alice like the wind, and was just in time to hear it say, as it turned a corner, "Oh my ears and whiskers, how late it's getting!" She was close behind it when she turned the corner, but the Rabbit was no longer to be seen: she found herself in a long, low hall, which was lit up by a row of lamps hanging from the roof.

There were doors all round the hall, but they were all locked; and when Alice had been all the way down one side and up the other, trying every door, she walked sadly down the middle, wondering how she was ever to get out again.

Suddenly she came upon a little three-legged table, all made of solid glass; there was nothing on it but a tiny golden key, and Alice's first idea was that this might belong to one of the doors of the hall; but, alas! either the locks were too large, or the key was too small, but at any rate it would not open any of them.

7.3. Worksheet 7b: The Quantifiers ALL and THERE EXISTS

The central topic of this chapter — the quantifiers ALL and THERE EXISTS — kicks off with Worksheet 7b, which begins with the excerpt shown above (Section 7.2), beginning with the sentence "There were doors all round the hall...".

Remarks:

- Worksheet 7b is to be completed by students in pairs or in small groups to allow for consultation.
- It is suggested that upon completion of the worksheet, students present their answers to the class, in order to initiate a whole class discussion of the solutions.
- At this point it is advised to avoid any judgment of students' answers; points of disagreement and misperceptions, however, should be noted.
- After presentation of the logical and mathematical background by the teacher (Section 7.4 below), students' answers should be revisited and discussed once again.

Worksheet 7b and Proposed Solutions

1. Lewis Carroll wrote:

"There were doors all round the hall, but they were all locked; and when Alice had been all the way down one side and up the other, trying every door...".

Below are several statements. Determine which of them describes a case that could contradict the condition in which "they were all locked." Underline "contradicts" or "does not contradict," accordingly. Explain your reasoning.

Note: The negation of "a locked door" is "an unlocked door." An unlocked door could be open or closed. In the context of this exercise, we consider only "an open door" as the negation of "a locked door."

Case A: Alice found all the doors to be open.

<u>Contradicts</u> / does not contradict, because: *It is given that all the doors were locked; therefore it could not be that all the doors were open.*

Case B: Alice found that there exists at least one open door.

<u>Contradicts</u> / does not contradict, because: *It is given that all the doors were locked; therefore it could not be that there exists even one door that is open.*

Case C: Alice found that all the doors were not locked.

<u>Contradicts</u> / does not contradict, because: *It is given that all the doors were locked.*

Case D: Alice found that not all the doors were locked.

<u>Contradicts</u> / does not contradict, because: *It is given that all the doors were locked.*

Case E: Alice found that not all the doors were open.

Contradicts / <u>does not contradict</u>, because: *In stating: "Not all the doors were open," it means that there exists at least one door that is not open; in other words: there exists at least one locked door. Yet the statement: "Not all the doors were open" does not mean that some were open. It is possible that all of them were locked. Therefore there is no contradiction.*

Case F: Alice found at least one door that was not open.

Contradicts / <u>does not contradict,</u> because: *In stating "At least one door was not open," or, in other words — there exists at least one door that is not open — this does not mean that some of the doors were open. It is possible that all were not open; that is — that all were locked. Therefore there is no contradiction.*

Case G: Alice found that all the doors were not open.

Contradicts / <u>does not contradict,</u> because: *From the statement "All the doors were not open," we may deduce the equivalent statement with certainty — "All the doors were locked."*

2.   a. Form a statement with the word ALL similar to the statement: "All parts of the little three-legged table were made of solid glass."

All residents of Wonderland wear blue shirts.

b. Form a statement that negates the statement you formed previously, but does not begin with the words "Not all."

In Wonderland there exists at least one resident who does not wear a blue shirt (similar to "There exists a part in the little three-legged table that is not made of solid glass").

7.4. Logical and Mathematical Background

Upon completion of Worksheet 7b, and after discussion of students' answers, the following concepts and topics are to be presented:

- A. Mathematical claims and their proofs
- B. The quantifiers ALL, THERE EXISTS, and ONLY
- C. Existence theorems and uniqueness theorems
- D. Quantifier representation using Venn diagrams
- E. Use of quantifier in day-to-day language

A. Mathematical claims and their proofs

Usually, a **mathematical claim** or **assertion** consists of two parts: (1) an assumption or assumptions (which typically appear after the word IF); (2) conclusion drawn from the assumptions (which typically appears after the word THEN). (Conditional statements of the form "IF... THEN..." are discussed in details in Chapter 8.)

For example:

- If two angles of a triangle have different measures, then the sides opposite those angles have different lengths, and opposite the larger angle is the longer side.
- If an integer is divisible by 4, then it is also divisible by 2.
- If the diagonals of a quadrilateral bisect each other and are perpendicular to one another, then the quadrilateral is a rhombus.
- If a number is a prime integer, then it is even only in one case.

What is a proof of a mathematical claim?

A proof of a claim is a sequence of arguments leading from the claim, through a series of deductions, to the conclusion, thus showing unequivocally that the conclusion follows from the assumptions. In this case the claim is said to be valid, meaning that the claim is a true statement.

A mathematical claim that has been proven is called a theorem. In other words, when a particular assertion is said to be a "theorem," it means that a proof of the claim is known, and the claim is valid without any doubt and with no exception, as opposed to a claim in court, for example, that only needs to be proven beyond reasonable doubt. There are a variety of methods for proving mathematical claims.

Examples:

- Proof by induction.
- Proof by negation.
- Proof by exhaustion. This is the method that Alice used to prove the assertion that all the doors in the hall were locked. She checked all the doors in the hall one at a time to verify it. This method may be used **only** when there are a **finite** number of cases to be examined.

B. The quantifiers ALL, THERE EXISTS, and ONLY

Proof of the statement: "In the Euclidean plane, the sum of interior angles of every triangle is 180°" focuses on the fact that the property holds for **all** triangles.

Proof of the statement: "The equation $2x^2 + 3x - 5 = 0$ has a solution in the integers" focuses on the **existence** of such a solution. (The equation does have a solution, it even has two: 1 and -2.5, one of which is, in fact, an integer.)

Proof of the statement: "There exists only one even prime number" focuses on the fact that **only** the number 2 has the above-mentioned property.

ALL, **THERE EXISTS**, and **ONLY** are types of assertions of quantity; they are therefore known as **quantifiers**.

In mathematical statements, quantifiers always appear together with a named set of mathematical objects, such as prime numbers, continuous functions, polygons, etc. In order to state that every real number satisfies a certain inequality, or that there exists a function with certain properties, or that only a certain polygon is equilateral, we need quantifiers. When we use quantifiers, the set in question must be indicated.

We expand on each of these three quantifiers below.

B1. The universal quantifier (denoted ∀)

An assertion that includes the word **ALL** is called a **universal assertion.**

Using the quantifier ALL, assertions may be formulated relating to a particular property of all elements that belong to a particular set, without exception.

Examples:

- ALL integers divisible by 4 are divisible by 2.
  The set: The set of integers divisible by 4.
  The property: divisible by 2.

- In every triangle in the plane, the sum of interior angles is 180°.
  The set: The set of triangles in the plane.
  The property: the sum of interior angles is 180°.

Note: When we say: "Every p is q," this does not necessarily mean that "Every q is p."

Examples:

- Every square is a quadrilateral. It does not follow from this that every quadrilateral is a square. And, in fact, it is well known that **not** every quadrilateral is a square (for example, the trapezoid is not a square).

- Every even number is the successor of an odd number. Every odd number is the successor of an even number, but neither of these two statements follows from the other. Each of these assertions must be proven separately and independently from the other.

As mentioned above, one of the methods of proving a universal assertion on a finite set is by exhausting all possibilities. In general, in order to prove a universal assertion, the rules of deduction that are used for proofs should be applied.

In order to negate a universal assertion of the form "every p is q," it suffices to provide (at least) one counterexample; that is, that there exists a p that is not q. Such a counterexample proves that the assertion is not universal, since it has (at least) one exception. Since the exception does not satisfy the assertion, the assertion is not universal.

Examples:

- The assertion "All prime numbers are odd" may be proven FALSE using the number 2 as a counterexample. In other words, THERE EXISTS a prime number that is NOT odd.

- The assertion "In each quadrilateral in the plane, the sum of interior angles is 360°"" may be proven FALSE by the quadrilateral with crossed sides which serves as a counterexample (as in the figure in Chapter 2, Section 2.6C (page 39)). In other words, THERE EXISTS a quadrilateral in the plane the interior angles sum of which is NOT equal to 360°.

B2. The existential quantifier THERE EXISTS (denoted ∃)

An assertion that contains the phrase **THERE EXISTS** (or the equivalent, such as "there is") is called an **existence assertion**. Using the quantifier THERE EXISTS, assertions may be formulated that indicate that the set in question has at least one member with the indicated property.

Examples:

- There exists a natural number divisible by 5 (divisible means, leaving no remainder).

  In mathematical notation: $\exists x \in N, \frac{x}{5} \in N$

  The set: the natural numbers.
  The property: divisible by 5.

  In the set of natural numbers there exists (at least) one number with this property (in fact there are many such numbers, e.g. 25, 110). It is worth noting that in this case the set also contains numbers that do not have the property (e.g. 13, 1743), although the existence assertion does not require this to be the case.

- There exists a quadrilateral in the plane with two opposite sides that are equal.

The set: the set of quadrilaterals in the plane.

The property: equality of two opposite sides.

In the set of quadrilaterals in the plane, there exist those that possess this property (e.g. parallelograms, isosceles trapezoids), and in such a case there also exist quadrilaterals in the set that do not possess the property (e.g. non-rhombic kites), although the existence assertion does not require this to be the case.

When proving: "There exists p that is q," the intent is that there exists at least one p that is q. Yet this does not mean that there also exists p that is not q. It is possible that every p is q, and it would still be correct to say that there exists p that is q.

Example:

• There exists a triangle in the plane the interior angles sum of which is 180°. There does not exist a triangle in the plane exceptional to this rule. All triangles possess that property.

Note: From a universal theorem (a proven claim) on a non-empty set there follows an existence theorem. For example, from the statement: "The sum of exterior angles of every convex polygon in the plane is 360°," it necessarily follows that there exists in the plane a convex polygon the exterior angles sum of which is 360°. But the opposite is not true; from an existence theorem one cannot deduce a universal assertion. For example, from the statement: "There exists an even number that is divisible by 6," it does not follow that every even number is divisible by 6.

Note: The negation of a universal assertion is equivalent to an existential assertion of a counterexample. That is, each follows from the other. For example, the assertion "**There exists** (at least) one number that is divisible by 5 that does not end in the digit zero" is equivalent to the assertion "**Not every** integer that is divisible by 5 ends in the digit zero." The assertion "**There exists** a human being who is flawless" is equivalent to the assertion that "**Not every** human being has flaws." This works the other way around as well; for example, the assertion: "**All** rivers flow to the sea" is equivalent to the assertion "**There does not exist** a river that doesn't flow to the sea."

Note: In order to prove an existential assertion, it suffices to identify a single member in the indicated set that satisfies the relevant property. Such a proof is called a **constructive proof of existence.** Sometimes, existence may be proven by proving the negation of the equivalent universal assertion. In this case the proof is not constructive.

Note: In order to disprove an existence assertion, it is necessary to prove the universal assertion equivalent to its negation by applying the established deductive reasoning process for proofs (details on this appear in Section 8.7f).

B3. The quantifier ONLY

Please note that there are logicians who do not consider ONLY to be a quantifier.[1] Although they view **ONLY** in a different category from THERE EXISTS and from ALL, it has been included in this chapter on quantifiers nonetheless.

The word ONLY is used in many different contexts.

For example:

- ONLY can be used as a quantifier, as in: (1) "Only people who like to play with numbers in their childhood grow up to be mathematicians" — that is, those who did not like to play with numbers in their childhood, did not grow up to be mathematicians. In other words, ALL those who grew up to be mathematicians liked to play with numbers in their childhood. (2) "Only numbers divisible by 2 are divisible by 6" — that is, if a number is not divisible by 2 then it is not divisible by 6. In other words, all numbers divisible by 6 are divisible by 2.

- ONLY can be used to limit other quantifiers, as in: (1) "There exists only one prime number that is even." In contrast to: "There exists one...," which implies "at least one" or "one or more;" the usage of ONLY in this case limits the existence to a single value. "all perfect squares, and only them, have an odd number of divisors." Indicating that ALL of the

---

[1] See, for example: Maier, E. (2006). *Belief in Context: Towards a Unified Semantics of de re and de se Attitude Reports.* Unpublished doctoral dissertation, Radboud University Nijmegen. Retrieved 2010 from: http://ncs.ruhosting.nl/emar/diss/.

perfect squares have this property, without stating ONLY them, implies that there could also be other numbers with that property.

- In spoken language ONLY can be used in the sense of "but"; for example: "I would head home now, only I don't have a ride."

- In spoken language ONLY can be used as the connective "only if" (which we did not address in the chapter on connectives) as an indicator of cause and effect. An example of this is: "The baseball game will be canceled only if it rains tomorrow." The difference between "The game will be canceled **if** it rains tomorrow" and "The game will be canceled **only if** it rains tomorrow," is that in the first case the game may be canceled even if it does not rain tomorrow — for example, if the lead player is injured — while in the latter case the game will not be canceled if something happens to one of the players, nor for any other reason other than rain. More details on use of ONLY as a connective appear in Chapter 8, which addresses conditional statements.

Notes:

- An assertion of the type "Only if q, then p" is equivalent to the assertion "If p then q." For example, the assertion "If a number is divisible by 6 then it is even" is equivalent to the assertion "Only if a number is even is it divisible by 6." Similarly, in spoken language, the assertion "If it is raining then it is cloudy" is equivalent to the assertion "Only if it is cloudy is it raining." This does not mean, of course, that there can't be clouds when it isn't raining. Notice that in the latter case the wording "Only if..." is easier to understand. In the mathematical example above, the clearer wording is the "If... then...".

- As we have seen with other examples, although there exists equivalence within each pair of assertions, there is a different emphasis in each, and sometimes these are interpreted as having different meanings. Drawing conclusions from conditional statements should be carried out with utmost care (details on "necessary conditions" and "sufficient conditions" can be found in Chapter 9).

- The two assertions: "Every criminal is bad" and "Only bad people are criminals" are equivalent. As a rule, every assertion of the form: "Only q is p" is equivalent to the assertion: "Every p is q." Nevertheless, there are many cases in which use of ONLY is harder to understand than use of EVERY. Yet there is also room for misunderstanding in wording that uses EVERY. For example, if we say "Everyone who knows math is smart," some people may misinterpret this as "Only people who know math are smart," instead of the equivalent "Only smart people know math."

C. Existence theorems and uniqueness theorems

In mathematics, if a given problem can be solved with at least one solution, we say that THERE EXISTS a solution to the problem. If it is known that there exists one solution, and it is not known how many additional solutions exist if there are any at all, we say that there exists AT LEAST one solution. AT LEAST ONE means that there exist one or more solutions. In general this expression is used when existence can be established, but it is not necessarily known if there are additional solutions, and if so, how many.

An **existence theorem** indicates existence of an object, such as a solution to a problem or equation. It does not need to indicate how many solutions exist. Nor does an existence theorem need to indicate a method for solving the problem or a solution. If an existence theorem includes a method to **indicate** at least one object that exists or a means to **construct** such an object, then we say that the proof is a **constructive proof of existence**. If the proof of existence consists of the arguments that there does in fact exist at least one object with the relevant property, but does not present a means to construct it and does not point to at least one such object, then the proof is called a **non-constructive proof of existence**. Anyhow, an existence theorem negates the possibility that there are no objects of the indicated type.

The following conversation pokes fun at the linguistic precision of logicians: An economist, a physicist and a logician are on a train to Scotland. When they cross the border into Scotland they see a brown cow from the window, standing parallel to their direction of travel.

"Look," says the economist, "The cows in Scotland are brown."
"No," says the physicist, "there are brown cows in Scotland."
"No," says the logician, "there exists in Scotland at least one cow, one of whose sides is brown."

In mathematics, we often search for the existence of a solution, and then we ask if it is **unique**.

A **uniqueness theorem** addresses uniqueness of a mathematical object, that is, that there exists at most one object that satisfies the given requirements. A uniqueness theorem negates the possibility of **more than** one such object and leaves open the possibility of existence and non-existence of that single object.

An existence and uniqueness theorem establishes that there exists exactly one object. The existence guarantees at least one object, while the uniqueness guarantees that there does not exist more than one.

D. Quantifier representation using Venn diagrams

In Section 5.4D of Chapter 5, we saw how Venn diagrams may be used to indicate logical relations between an object and a set, as well as relations among sets. Statements that contain quantifiers may be translated to equivalent statements on sets; the properties can then be represented using Venn diagrams.

D1. Representation of universal statements using Venn diagrams

Let us examine the universally true statement "Every square is a quadrilateral." The meaning of this statement is that there does not exist a square that is not a quadrilateral. In other words, the set of squares is included in the set of quadrilaterals; in fact, it is a proper subset of that set. The corresponding Venn diagram looks like this:

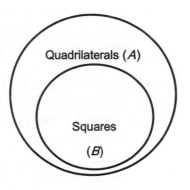

As a rule, this describes the universal statement:
For all $x \in B$, $x \in A$.

D2. Representation of existence statements using Venn diagrams

Let us examine the existence statement (which is true): "In the set of even numbers there exists a number that is divisible by 3." The meaning of this statement is that in the set of even numbers there exists a non-empty subset (or a proper subset) whose elements are numbers that are divisible by 3. The corresponding Venn diagram is shown in Figure A marked by a solid line.

As a rule, this describes the existence statement: There exists an $x \in A$ such that $x \in B$.

Notes:

- The aforementioned does not preclude existence of other numbers (odd numbers) that are divisible by 3, nor does it require existence of these other numbers.

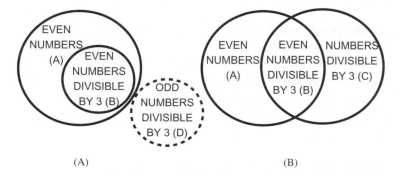

(A)   (B)

- Our statement establishes that there is a non-empty intersection between the set of numbers divisible by 3 and the set of even numbers. This is shown in the Venn diagram in Figure B.

- The generally agreed upon definition of the trapezoid is "a quadrilateral with exactly one pair of parallel sides." Let us examine the statement "There does not exist a trapezoid all angles of which are equal." This statement (which is true), negates the existence of trapezoids with the above mentioned property. The meaning of this statement is that the set of trapezoids is disjoint from the set of polygons all angles of which are equal. That is, the intersection of the two sets is the empty set. The corresponding Venn diagram looks as follows:

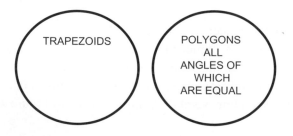

D3. Representation of statements that include the quantifier ONLY using Venn diagrams

Let us examine the following statement (which is true): "Only quadrilaterals are squares." The meaning of this statement is that there does not exist a square that is not a quadrilateral. In other words, all squares are quadrilaterals (similar to Section D1 above). The figure on the right demonstrates the corresponding Venn diagram. As a rule, this describes the statement: Only $x \in A$ is $x \in B$ (which is equivalent to: For all $x \in B, x \in A$).

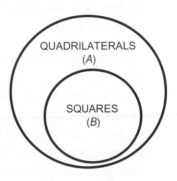

E. Use of quantifiers in day-to-day language

Expressions containing universal assertions and existence assertions are used in day-to-day language without ever giving them a second thought. For example, the phrase "Everyone knows that ..." does not mean that in fact there does not exist someone who does not know that fact. Similarly, a statement such as "There are people with hearts of stone" establishes something physiologically false. In these cases, the use of quantifiers is intended to emphasize feelings or to reinforce metaphors (borrowed expressions).

7.5. Review of Answers to Worksheet 7b

After students have become familiar with the logical background, it is advised to ask them to go back and review their answers to Worksheet 7b and correct them if necessary.

7.6. A Look Inside the Text

Read the excerpt from page 7 of *Alice's Adventures in Wonderland* that appears in boldface below. This excerpt is the continuation of the one you read above. It will serve as the basis for further discussion of the quantifiers ALL and THERE EXISTS.

However, on the second time round, she came upon a low curtain she had not noticed before, and behind it was a little door about fifteen inches high: she tried the little golden key in the lock, and to her great delight it fitted!

Alice opened the door and found that it led into a small passage, not much larger than a rat-hole: she knelt down and looked along the passage into the loveliest garden you ever saw. How she longed to get out of that dark hall, and wander about among those beds of bright flowers and those cool fountains, but she could not even get her head through the doorway; "and even if my head would go through," thought poor Alice, "it would be of very little use without my shoulders. Oh, how I wish I could shut up like a telescope! I think I could, if I only knew how to begin." For, you see, so many out-of-the-way things had happened lately, that Alice had begun to think that very few things indeed were really impossible.

7.7. Worksheet 7c: More on the Quantifiers ALL and THERE EXISTS

We continue with the central topic of this chapter — the quantifiers ALL and THERE EXISTS — in Worksheet 7c, which begins with the excerpt shown above (Section 7.6).

Remarks:

- This worksheet may be completed as individual practice, in pairs or in small groups, with group discussion encouraged.
- Upon completion of the worksheet, it is recommended that the students present their answers to the class to initiate a whole class discussion of the solutions.

Worksheet 7c and Proposed Solutions

Alice Finds a Door

However, on the second time round, she came upon a low curtain
she had not noticed before, and behind it was a
little door about fifteen inches high:
she tried the little golden key in the lock,
and to her great delight it fitted! (page 7)

1. This paragraph recounts that Alice found evidence contradicting an assertion that appears in Worksheet 7b. What is the evidence that Alice found, and what assertion does it contradict?

    Alice found a door that could be opened with the golden key. This contradicts the assertion that the key does not open any of the doors in the hall.

2. The people of HAPPYLAND claim that EVERY little door is painted red. Alice came to HAPPYLAND and found evidence contradicting this assertion. What did Alice find?

    Evidence contradicting the assertion: "Every little door is painted red" would be a little door that is not painted red. One cannot tell whether any little door like that is painted, and if so what color it is painted with.

3. The people of MAGICLAND claim that EVERY door that is painted red is little. Alice arrived at MAGICLAND and found evidence contradicting this assertion. What did Alice find?

    Evidence contradicting the assertion "Every door that is painted red is little" would be a red painted door that is not little. Because "not little" is "big" it would be correct to use "big" instead of "not little."

4. If the people of HAPPYLAND were to claim that ONLY little doors are painted red, would the evidence you thought Alice found in your answer to Question 2 contradict this claim as well?

    The answer is "No." In question 2 the evidence contradicting the assertion "Every little door is painted red" was a little door that is not painted red. Evidence that contradicts the assertion "Only little doors are painted red," however, would be a large door painted red. Finding a little door that is not painted red, therefore, does not contradict the assertion in this question.

5. If the people of MAGICLAND were to claim that only doors painted red are little, would the evidence you thought Alice found in your answer to question 3 contradict this claim as well?

    The answer is "No." In question 3 the evidence contradicting the assertion "Every door that is painted red is little" was a red painted door that is not little. Evidence that contradicts the assertion "Only doors painted red are little," however, would be a little door that is not painted red. Finding a door that is not little and is painted red, therefore, does not contradict the assertion in this question.

7.8. Worksheet 7d: The Quantifiers ALL and ONLY

Worksheet 7d addresses the quantifiers ALL and ONLY.

Remarks:

- This worksheet may be performed as individual practice, in pairs or in small groups, with group discussion encouraged.
- Upon completion of the worksheet, it is recommended that the students present their answers to the class to initiate a whole class discussion of the solutions.

Worksheet 7d and Proposed Solutions

In the course of her wanderings in Wonderland, Alice met a Cheshire Cat. Alice chatted with it about the people who live in Wonderland. Below is an excerpt from the conversation between them (page 76):

"In *that* direction," the Cat said, waving its right paw round, "lives a Hatter: and in that direction," waving the other paw, "lives a March Hare. Visit either you like: they're both mad."

"But I don't want to go among mad people," Alice remarked.

"Oh, you can't help that," said the Cat: "we're all mad here. I'm mad. You're mad."

"How do you know I'm mad?" said Alice.

"You must be," said the Cat, "or you wouldn't have come here."

1. Each row in the table below contains two statements followed by a conclusion. Examine the conclusion with regard to the two statements and underline one of the options: True, False, Cannot be determined (due to insufficient information). Explain your reasoning.

    A Venn diagram may be helpful in solving the problem.

    As indicated in Section 7.4d, the statement "For every x in A, x is in B" may be represented by the Venn diagram below:

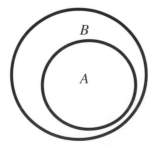

This diagram also represents the statement: "Only x∈B is x∈A". Therefore, we represent the statement: "All those who come to

Wonderland are mad" by Diagram A, and the statement "Only those who come to Wonderland are mad" by Diagram B.

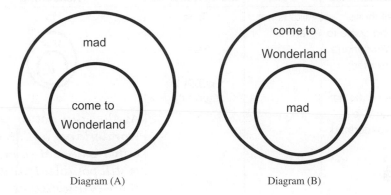

Diagram (A)   Diagram (B)

Let us use these diagrams to find the solution. Diagram A above corresponds to the statements that appear in the odd entries of the table, while Diagram B above corresponds to the even ones.

Let us designate the set of those who come to Wonderland by E, and the set of mad creatures by M.

Let us designate Alice by a.

Row	The Statements	Conclu-sions	Conclusions Drawn from the Statements	Reasoning
1	All those who come to Wonderland are mad; Alice did not come to Wonderland.	Alice is not mad.	True False <u>Cannot be determined</u>	$\overline{M}$ .a M .a E *As Alice did not come to Wonderland, $a \notin E$. But either case $a \in M - E$ or $a \in \overline{M}$ is still possible. Therefore, there is insufficient information given to determine whether or not Alice is mad.*
2	Only those who come to Wonderland are mad; Alice came to Wonderland.	Alice is mad.	True False <u>Cannot be determined</u>	$\overline{E}$ E .a .a M *As Alice did come to Wonderland, $a \in E$. But either case $a \in M$ or $a \in E - M$ is still possible. Therefore, there is insufficient information given to determine whether or not Alice is mad.*

(Continued)

(*Continued*)

Row	The Statements	Conclu-sions	Conclusions Drawn from the Statements	Reasoning
3	All those who come to Wonderland are mad; Alice came to Wonderland.	Alice is not mad.	True <u>False</u> Cannot be determined	*As Alice did come to Wonderland, $a \in E$. Since it is true that $E \subset M$, it follows that $a \in M$ is true as well. That is, the conclusion is that Alice is mad, and therefore it cannot be said that Alice is not mad.*
4	Only those who come to Wonderland are mad; Alice did not come to Wonderland.	Alice is mad.	True <u>False</u> Cannot be determined	*As Alice did not come to Wonderland, $a \notin E$. Since $M \subset E$ is true, it follows that $a \notin M$ is true as well. That is, the conclusion is that Alice is not mad, therefore it cannot be said that Alice is mad.*

(*Continued*)

(Continued)

Row	The Statements	Conclu-sions	Conclusions Drawn from the Statements	Reasoning
5	All those who come to Wonderland are mad; Alice did not come to Wonderland.	Alice is mad.	True False <u>Cannot be determined</u>	 *As Alice did not come to Wonderland, $a \notin E$. But either $a \in M - E$ or $a \in \overline{M}$ are still likely possibilities. Therefore there is insufficient information given to determine whether or not Alice is mad.*
6	Only those who come to Wonderland are mad; Alice came to Wonderland.	Alice is not mad.	True False <u>Cannot be determined</u>	 *As Alice did come to Wonderland, $a \in E$. But either $a \in M$ or $a \in E - M$ are still likely possibilities. Therefore there is insufficient information given to determine whether or not Alice is mad.*

(Continued)

(*Continued*)

Row	The Statements	Conclu-sions	Conclusions Drawn from the Statements	Reasoning
7	All those who come to Wonderland are mad; Alice came to Wonderland.	Alice is mad.	<u>True</u> False Cannot be determined	*As Alice did come to Wonderland, a ∈ E. Since E ⊂ M, it follows that a ∈ M that is Alice is mad.*
8	Only those who come to Wonderland are mad; Alice did not come to Wonderland.	Alice is not mad.	<u>True</u> False Cannot be determined	*As Alice did not come to Wonderland, a ∈ E̅. Since M ⊂ E is true, it follows that a ∉ M is true as well. That is, we conclude that Alice is not mad.*

2. Which of the eight rows in the table corresponds to the description in the excerpt at the top of this worksheet?

The corresponding row in the table is row 7.

7.9. Worksheet 7e: A Group Exercise for Quantifiers

Worksheet 7e will help student summarize the understanding they have acquired on quantifiers.

Remarks:

- This worksheet is intended to be a group exercise.
- Upon completion of the worksheet, it is recommended that the students present their answers to the class to initiate a whole class discussion of the solutions.

Worksheet 7e: A Group Exercise for Quantifiers

Write an imaginary dialogue between the White Rabbit and Alice in which the Rabbit tries to explain to Alice the difference between the statements:

"All *a* are *b*," "There exists *a* which is *b*," "Only *a* is *b*" using examples, as well as by examples of statements negating these statements.

It is advised that the students take time out to read the dialogues aloud.

Alice, today we are going to learn about the differences among statements containing ALL, THERE EXISTS, and ONLY.

ONLY you could explain ALL these to me....

7.10. Worksheets 7f and 7g: Summary Exercise for the Quantifiers ALL, THERE EXISTS, and ONLY

The summary exercise is composed of two worksheets: 7f and 7g.

Remarks:

- Worksheet 7f is to be performed as individual practice by students. In Worksheet 7g (which appears in the students' workbook only) students will compare their current answers with the answers they gave at the start of the chapter (Worksheet 7a).
- Upon completion of the summary exercise it is recommended that a whole class discussion be held in order to consider the changes that have taken place in students' understanding and perceptions through the course of this chapter, as well as the particular difficulties encountered with the subject matter.

Worksheet 7f and Proposed Solutions

Answer the questions in this worksheet in their entirety, providing as much detail as possible. If necessary, you may indicate: "I did not understand the question, therefore I have not answered it. Make sure not to go back to Worksheet 4a before you complete your work on this worksheet.

In each of the four sections presented below, there is a true statement followed by a table. In the table accompanying each statement, the first column includes ten possible conclusions that may follow from it. The headings of the three additional columns are: "it follows from the statement that the conclusion is true," "it does not necessarily follow from the statement that the conclusion is true," "it follows from the statement that the conclusion is false." Choose the column corresponding to the right answer for each conclusion, mark it by "+" and explain why you think it is correct.

1. The true statement: **Of all the creatures in Wonderland, only White Rabbits have a waistcoat pocket.**

	Conclusion	It follows from the statement that the conclusion is true	It does not necessarily follow from the statement that the conclusion is true	It follows from the statement that the conclusion is false
A	In Wonderland there does not exist a Red Rabbit with a waistcoat pocket.	*From the statement: "Only White Rabbits have waistcoat pockets" it follows that every other creature (even a Rabbit that is not white) has no waistcoat pocket. The conclusion is therefore true.*		
B	In Wonderland there does not exist a White Rabbit with a pants' pocket.		*The given statement relates to a waistcoat pocket. No conclusions may be drawn regarding a pants' pocket. It is possible that White Rabbits have pockets also in their pants.*	

(Continued)

(*Continued*)

	Conclusion	It follows from the statement that the conclusion is true	It does not necessarily follow from the statement that the conclusion is true	It follows from the statement that the conclusion is false
C	In Wonderland there exists a Green Rabbit with a waistcoat pocket.			*It is given that only White Rabbits have a waistcoat pocket; from this it follows that every other creature (even a Rabbit, if it is not white) has no waistcoat pocket.*
D	In Wonderland there exists a white cat with a waistcoat pocket.			*Only White Rabbits have a waistcoat pocket; that is -- every other creature has no waistcoat pocket.*
E	In Wonderland every creature that is not a White Rabbit has no waistcoat pocket.	*The meaning of the statement "Only White Rabbits have waistcoat pockets" is that every other creature has no waistcoat pocket.*		

(*Continued*)

(*Continued*)

	Conclusion	It follows from the statement that the conclusion is true	It does not necessarily follow from the statement that the conclusion is true	It follows from the statement that the conclusion is false
F	Robbie, the youngest of the White Rabbits in Wonderland, has a waistcoat pocket.		*It is known that only White Rabbits have a waistcoat pocket, but it does not follow from this that every White Rabbit has a waistcoat pocket. It is possible that there are those that do not have such a pocket; therefore it could be that Robbie is one of those that does not have one, but it could also be that it does.*	
G	In Wonderland, every White Rabbit has a waistcoat pocket.		*It is given that only White Rabbits have a waistcoat pocket, but it does not necessarily follow from this that every White Rabbit has a waistcoat pocket. Perhaps it does, and perhaps not.*	

(*Continued*)

		(Continued)		
	Conclusion	It follows from the statement that the conclusion is true	It does not necessarily follow from the statement that the conclusion is true	It follows from the statement that the conclusion is false
H	In SURPRISELAND, only White Rabbits have a waistcoat pocket.		*In the given statement there is information regarding White Rabbits in Wonderland. Nothing is known about White Rabbits in other lands. It could be that they have waistcoat pockets, and it could be that they don't.*	
I	In Wonderland, there exists a White Rabbit that has waist-coat buttons.		*The given statement does not relate to existence or non-existence of White Rabbits in Wonderland that have waistcoat buttons. Maybe they do and maybe they don't.*	
J	Barney the Pink Rabbit in Wonderland has no waistcoat pocket.	*Barney is indeed a rabbit, and it lives in Wonderland, but is Pink, and in Wonderland only White Rabbits have waistcoat pockets.*		

2.   The true statement: **Every little girl named Alice likes books with pictures in them.**

	Conclusion	It follows from the statement that the conclusion is true	It does not necessarily follow from the statement that the conclusion is true	It follows from the statement that the conclusion is false
A	Alice Smith is a little girl who does not like books with pictures in them.			*Alice Smith is a little girl named Alice. Since every little girl named Alice likes books with pictures in them, it follows that Alice Smith likes books with pictures in them.*
B	Every little girl named Alice does not like newspapers with pictures in them.		*The statement says that every little girl named Alice likes books with pictures in them. It cannot be determined whether or not little girls named Alice also love **newspapers** with pictures in them.*	
C	Grandma Alice does not like books with pictures in them.		*The statement is about little girls named Alice. It cannot be determined whether or not grandmothers named Alice like books with pictures in them.*	

(Continued)

(*Continued*)

	Conclusion	It follows from the statement that the conclusion is true	It does not necessarily follow from the statement that the conclusion is true	It follows from the statement that the conclusion is false
D	Every little girl named Alice likes books that have no pictures in them.		*The statement says that every little girl named Alice likes books with pictures in them. It does not say anything about how little girls named Alice feel about books that have no pictures in them. Perhaps they like them, and perhaps they don't.*	
E	In SURPRISELAND, there exists a little girl named Alice that does not like books with pictures in them.			*The statement says that every little girl named Alice, regardless of where she lives likes books with pictures in them. Therefore it can't be true that there exists a little girl in Surpriseland that **does not** like books with pictures in them.*

(*Continued*)

(Continued)

	Conclusion	It follows from the statement that the conclusion is true	It does not necessarily follow from the statement that the conclusion is true	It follows from the statement that the conclusion is false
F	There does not exist in the world a little girl named Alice that does not like books with pictures in them.	*Every little girl named Alice likes books with pictures in them; so it can't be the case that there exists a little girl named Alice who does not like books with pictures in them. The conclusion is therefore true.*		
G	Dinah is a little girl who likes books with pictures in them.		*The statement is about little girls named Alice. It says nothing about girls named Dinah. It cannot be determined whether or not little girls named Dinah like books that have no pictures in them.*	
H	Every little girl whose name is not Alice does not like books with pictures in them.		*The statement is about little girls named Alice. From it, nothing can be determined about little girls with any other name.*	

(Continued)

(Continued)

	Conclusion	It follows from the statement that the conclusion is true	It does not necessarily follow from the statement that the conclusion is true	It follows from the statement that the conclusion is false
I	Not every little girl named Alice likes books with pictures in them.			*It is given that every little girl named Alice **does** like books with pictures in them. The conclusion for this question contradicts the given statement.*
J	Alice Robertson is a little girl who likes books with pictures in them.	*It is given that every little girl named Alice likes books with pictures in them. Alice Robertson is a little girl named Alice; therefore it follows that the statement applies to her (as well), and establishes that she likes books with pictures in them.*		

3. The true statement: **In Wonderland there exist Rabbits who can talk.**

	Conclusion	It follows from the statement that the conclusion is true	It does not necessarily follow from the statement that the conclusion is true	It follows from the statement that the conclusion is false
A	In Wonderland there exists a Rabbit that can't talk.		*According to the statement, it is known that in Wonderland there are Rabbits who **can** talk. It is possible that all the Rabbits in Wonderland can talk, but it is also possible that this is not the case. Therefore it cannot be deduced that there also exist Rabbits that **cannot** talk, and it also cannot be deduced that such Rabbits don't exist.*	
B	In Wonderland there do not exist Rabbits that can sing.		*The given statement does not provide information about existence or non-existence of singing Rabbits. These may exist, but it may also be the case that they do not exist.*	

(Continued)

(Continued)

	Conclusion	It follows from the statement that the conclusion is true	It does not necessarily follow from the statement that the conclusion is true	It follows from the statement that the conclusion is false
C	In Wonderland there does not exist even one Rabbit that can't talk.		*It is known that in Wonderland there exist Rabbits who can talk. It is possible that **all** Rabbits in Wonderland can talk, but it is also possible that there also exist Rabbits who **can't** talk.*	
D	In Wonderland every Rabbit can talk.		*Indeed it is known that in Wonderland **there exist** Rabbits who can talk. But it is also possible that **not all** can talk, even though the opposite may be true as well.*	
E	In Wonderland there does not exist a Rabbit that can't talk.		*It is known that in Wonderland there exist Rabbits who can talk. It is not known whether there also exist Rabbits who can't talk.*	

(Continued)

(*Continued*)

	Conclusion	It follows from the statement that the conclusion is true	It does not necessarily follow from the statement that the conclusion is true	It follows from the statement that the conclusion is false
F	In Wonderland every creature that is not a Rabbit can't talk.		*The statement doesn't provide any information about creatures that are not Rabbits. Therefore nothing can be deduced from it about other creatures.*	
G	In Wonderland there exists at least one Rabbit who can talk.	*It is given that in Wonderland there exist Rabbits who can talk. This means that there exists at least one Rabbit of that kind.*		
H	In Wonderland there does not exist even one Rabbit who can talk.			*The given statement states the opposite — that in Wonderland* **there exist** *Rabbits who can talk. The conclusion therefore contradicts the given statement.*

(*Continued*)

(*Continued*)

	Conclusion	It follows from the statement that the conclusion is true	It does not necessarily follow from the statement that the conclusion is true	It follows from the statement that the conclusion is false
I	In SURPRISELAND there exist Rabbits who can talk.		*The statement is about Wonderland only. No information is given about Surpriseland or any other land.*	
J	In Wonderland there exist cats who can talk.		*The statement is about **Rabbits** in Wonderland. It doesn't provide any information about other creatures in Wonderland.*	

4.   The true statement: **There do not exist cats that do not eat carrot.**

	Conclusion	It follows from the statement that the conclusion is true	It does not necessarily follow from the statement that the conclusion is true	It follows from the statement that the conclusion is false
A	There exist cats that eat carrots.	*According to the given statement, there do not exist cats who do not eat carrots. This means that every cat eats carrots; therefore, it may certainly be deduced that there exist cats who eat carrots.*		

(*Continued*)

(*Continued*)

	Conclusion	It follows from the statement that the conclusion is true	It does not necessarily follow from the statement that the conclusion is true	It follows from the statement that the conclusion is false
B	There exist cats that do not eat carrots.			*The given statement states quite the opposite — that there **do not exist** cats like those mentioned in the conclusion. The conclusion therefore contradicts the given statement.*
C	There exist dogs that eat carrots.		*The statement is about **cats** and says nothing about dogs, therefore nothing can be concluded from it about dogs that eat carrots or don't.*	
D	All cats eat carrots.	*The conclusion is equivalent to the given statement. If there do not exist cats that do not eat carrots, it means that all cats **do** eat carrots.*		

(*Continued*)

(*Continued*)

	Conclusion	It follows from the statement that the conclusion is true	It does not necessarily follow from the statement that the conclusion is true	It follows from the statement that the conclusion is false
E	There do not exist cats that eat mice.		*The statement is about eating carrots and says nothing about eating mice, therefore no conclusions can be drawn from it about existence or non-existence of cats that eat mice.*	
F	There exist cats that eat bread.		*The statement is about eating carrots, therefore no conclusions can be drawn about existence or non-existence of cats that eat bread.*	
G	There exist cats that do not eat cabbage.		*The statement is about eating carrots, therefore no conclusions can be drawn about existence of non-existence of cats that eat cabbage.*	

(*Continued*)

	Conclusion	It follows from the statement that the conclusion is true	It does not necessarily follow from the statement that the conclusion is true	It follows from the statement that the conclusion is false
H	All cats drink milk.		The statement is about eating carrots, therefore no conclusions can be drawn about existence of non-existence of cats that drink milk.	
I	Alley cats do not eat carrots.			It can be derived from the given statement that all cats eat carrots. Since alley cats are cats, they too eat carrots; the conclusion, therefore, contradicts the given statement.
J	Kitty the cat eats carrots.	According to the given statement there do not exist cats that do not eat carrots. From this it follows that every cat eats carrots; therefore it certainly may be deduced that Kitty, who is a cat, eats carrots.		

(Continued)

Chapter 8

The Logical Connective
'IF…, THEN…' (Implication)

IF Alice peeks behind
the curtain,
THEN she will find the
door that leads to the
garden

8.1. Worksheet 8a: Opening Exercise. Lead-in to the Logical Connective 'IF...THEN'

Instruction of this chapter begins with Worksheet 8a Opening Exercise (here forth abbr. WS8a), to be completed as individual practice by each student. Upon completion of the task notify the students that they will return to this worksheet at the end of the chapter (Worksheet 8f Summary Exercise); then they will be able to compare their initial understanding with that which they acquired during the course of the chapter, while reflecting upon the changes in their understanding along with the points where further clarification is still required (Worksheet 8g, which appears only in the Student's Workbook). For this reason it is recommended that discussion of this worksheet will be held off until the end of the chapter. In some cases the teacher may prefer to collect WS8a and give it back only upon completion of WS8f when students are ready to work on WS8g.

Because this lead-in worksheet is repeated again as a summary exercise, the sheet (together with proposed solutions) appears only once — at the end of this chapter.

8.2. A Look Inside the Text

Read the two paragraphs from pages 7 and 8 of *Alice's Adventures in Wonderland* that appear in boldface below. This excerpt will serve as the basis for a scientific discussion of the telescope as well as a discussion of the mathematical concept of "proportion."

> Suddenly she came upon a little three-legged table, all made of solid glass; there was nothing on it but a tiny golden key, and Alice's first idea was that this might belong to one of the doors of the hall; but, alas! either the locks were too large, or the key was too small, but at any rate it would not open any of them. However, on the second time round, she came upon a low curtain she had not noticed before, and behind it was a little door about fifteen inches high: she tried the little golden key in the lock, and to her great delight it fitted!
>
> **Alice opened the door and found that it led into a small passage, not much larger than a rat-hole: she knelt down and looked along the passage into the loveliest garden you ever saw. How she longed**

to get out of that dark hall, and wander about among those beds of bright flowers and those cool fountains, but she could not even get her head through the doorway; "and even if my head would go through," thought poor Alice, "it would be of very little use without my shoulders. Oh, how I wish I could shut up like a telescope! I think I could, if I only knew how to begin." For, you see, so many out-of-the-way things had happened lately, that Alice had begun to think that very few things indeed were really impossible.

There seemed to be no use in waiting by the little door, so she went back to the table, half hoping she might find another key on it, or at any rate a book of rules for shutting people up like telescopes: this time she found a little bottle on it ("which certainly was not here before," said Alice,) and tied round the neck of the bottle was a paper label, with the words "DRINK ME" beautifully printed on it in large letters.

It was all very well to say "Drink me," but the wise little Alice was not going to do *that* in a hurry. "No, I'll look first," she said, "and see whether it's marked '*poison*' or not;" for she had read several nice little stories about children who had got burnt.

8.3. Scientific Discussion: The Telescope

To expand students' knowledge, it is advised to take the time for a closer look at this excerpt and hold a discussion with students on various scientific aspects of the telescope and how it operates. As there is no worksheet for this topic, the excerpt in the Student's Workbook contains the excerpt that appears in Section 8.2 together with the one that appears in Section 8.5. It is up to the teacher to advise students which part to read at what stage.

The telescope is an instrument used to observe far-away objects, such as stars. It enables seeing details that cannot be seen by the naked eye.

Students may be encouraged to search for additional information about the telescope and to address questions such as these:

- When was the telescope invented and by whom?
- What kinds of telescopes exist today?
- How do the different telescopes work?
- Who uses telescopes?

8.4. Discussion: The Mathematical Concept of "Proportion"

The excerpt above may serve as a basis for discussion of the meaning of "proportion" and "proportional expansion (or reduction)."

Proportion in mathematics is the equivalence between numerical ratios $a:b = c:d$ where a, b, c, d are real numbers, and b, d are non-zero.

Examples of mathematical problems that require considerations of proportion to solve may be found in textbooks on geometry (in particular with respect to similar triangles), algebra and trigonometry.

8.5. A Look Inside the Text

Read the two paragraphs from pages 8 and 9 of *Alice's Adventures in Wonderland* that appear in boldface below. They will serve as the basis for discussion of the logical connective 'IF ..., THEN ...'.

There seemed to be no use in waiting by the little door, so she went back to the table, half hoping she might find another key on it, or at any rate a book of rules for shutting people up like telescopes: this time she found a little bottle on it ("which certainly was not here before," said Alice,) and tied round the neck of the bottle was a paper label, with the words "DRINK ME" beautifully printed on it in large letters.

It was all very well to say "Drink me," but the wise little Alice was not going to do *that* in a hurry. "No, I'll look first," she said, "and see whether it's marked *'poison'* or not;" for she had read several nice little stories about children who had got burnt, and eaten up by wild beasts, and other unpleasant things, all because they *would* not remember the simple rules their friends had taught them: such as, that a red-hot poker will burn you if you hold it too long; and that, if you cut your finger *very* deeply with a knife, it usually bleeds; and she had never forgotten that, if you drink much from a bottle marked "poison," it is almost certain to disagree with you, sooner or later.

However, this bottle was *not* marked "poison," so Alice ventured to taste it, and finding it very nice (it had, in fact, a sort of mixed flavour of cherry-tart, custard, pineapple, roast turkey, coffee, and hot buttered toast,) she very soon finished it off.

"What a curious feeling!" said Alice. "I must be shutting up like a telescope."

And so it was indeed: she was now only ten inches high, and her face brightened up at the thought that she was now the right size for going through that little door into that lovely garden. First, however, she waited for a few minutes to see if she was going to shrink any further: she felt a little nervous about this: "for it might end, you know," said Alice to herself, "in my going out altogether, like a candle. I wonder what I should be like then?" And she tried to fancy what the flame of a candle looks like after the candle is blown out, for she could not remember ever having seen such a thing.

After a while, finding that nothing more happened, she decided on going into the garden at once; but, alas for poor Alice! when she got to the door, she found she had forgotten the little golden key, and when she went back to the table for it, she found she could not possibly reach it: she could see it quite plainly through the glass, and she tried her best to climb up one of the legs of the table, but it was too slippery; and when she had tired herself out with trying, the poor little thing sat down and cried.

8.6. Worksheet 8b: The Logical Connective 'IF... THEN...'

The central topic of this chapter — the logical connective 'IF ..., THEN ...' — kicks off with Worksheet 8b, which begins with the excerpt shown above (Section 8.5). As there is no worksheet for the previous topics, the excerpt in the Student's Workbook contains the excerpt that appears in Section 8.2 together with the one that appears in Section 8.5. It is up to the teacher to advise students which part to read at what stage.

Remarks:

- Worksheet 8b is to be completed in pairs or in small groups to allow for consultation.
- It is suggested that upon completion of the worksheet, students present their answers to the class, in order to initiate a whole class discussion of the solutions.
- At this point it is advised to avoid any judgment of students' answers; points of disagreement and misperceptions, however, should be noted.
- After presentation of the logical and mathematical background by the teacher (Section 8.7 below), students' answers should be revisited and discussed once again.

Worksheet 8b and Proposed Solutions

1. Find as many sentences as possible in the text that can be reworded such that they fit the format 'IF ..., THEN ...', and reword them as such.

    A. *If you hold a red-hot poker too long, then it will burn you.*

    B. *If you cut your finger very deeply with a knife, then it usually bleeds.*

    C. *If you drink much from a bottle marked "poison," then it is almost certain to disagree with you, sooner or later.*

2. Can one tell with certainty that the bottle Alice drank from did not contain poison? Why?

    It cannot be determined with certainty whether the bottle Alice drank from did not contain poison. There may be poison in the bottle even if it is not marked "poison." If the bottle was marked "poison" (assuming that the label is reliable), then we would know with certainty that it contained poison. But if the bottle is unmarked, we cannot determine what is in it.

3. How could Alice know with certainty that the bottle did not contain poison? Why?

    Only by drinking the contents of the bottle and examining its effect (don't try this at home!) could Alice know with certainty whether or not the bottle contained poison. There is no other way to verify this without laboratory access or some other means of verification. (Alice was smart enough not to do it...)

8.7. Logical and Mathematical Background

Upon completion of Worksheet 8b, and after discussion of students' answers, the following concepts and topics are to be presented:

  A. Conditional statements
  B. How to negate a conditional statement
  C. What can be inferred from a conditional statement
  D. What cannot be inferred from a conditional statement
  E. The truth table for the implication connective
  F. Rules of inference
  G. Logical fallacies in drawing conclusions
  H. Bi-directional conditional statements

A. Conditional statements

Given two statements, two (typically different) conditional statements may be constructed from them by using the logical connective 'IF ..., THEN ...' as follows: Write the word IF followed by one of the statements, then add a comma and write the word THEN followed by the second statement.

For example:

- Statement 1: It is raining in Wonderland.
  Statement 2: The White Rabbit is wet.

One of the two conditional statements that may be obtained is "If it is raining in Wonderland, then the White Rabbit is wet." The first portion — the protasis of the conditional statement — which appears after the word IF ("it is raining in Wonderland"), is called the **antecedent**. The second portion — the apodosis — which appears after the word THEN ("the White Rabbit is wet"), is called the **consequent**. The connective 'IF ..., THEN ...' is called the **implication connective.** The second conditional statement that can be constructed from these two is in the reverse order: "If the White Rabbit is wet, then it is raining in Wonderland."

Notes:

- The conditional statement that can be derived may be meaningless or illogical. For example, the statement "If I buy shoes, then I will paint my car red" is not likely to be encountered in reality. Despite this, everything stated below about conditional statements applies to conditional statements of that nature as well.

- Even if we construct a conditional statement from two true mathematical statements, the result may be meaningless if there is no connection between the two components. An example of this is the statement "If two is a prime number, then all angles in a rectangle are 90°." Despite this, everything stated below about conditional statements applies to conditional statements of that nature as well.

- The meaning of a conditional statement relates, for the most part, to the existence of the consequent that follows from the assumption indicated in the antecedent. Yet it is important to keep in mind that logic deals with the structure of statements and the ramifications therein, and not with their content.

- If we denote the antecedent by p and the consequent by q, then the conditional statement "if p, then q" is denoted as follows: $p \rightarrow q$. The statement can also be expressed in other ways, such as "p implies q" and "q follows from p".

- As indicated above, the conditional statement $p \rightarrow q$ has two parts — the antecedent (p) and the consequent (q) — each of which is a statement in itself. The two statements that compose the conditional statement could themselves be statements that consist of connectives or quantifiers (the OR here refers, of course, to Inclusive OR). Thus, for example, the connective NOT may appear on either part. As a result we have four possible cases:

  1. p and q do not include negation ("If it is raining in Wonderland, then the White Rabbit is wet").
  2. p includes negation and q does not ("If Alice does not find a key, then she will remain in the hall").
  3. p does not include negation and q does ("If cats eat bats, then bats do not eat cats").
  4. p and q both include negation ("If Alice had not seen the White Rabbit, then she would not have entered the tunnel").

- Since the connective 'IF ..., THEN ...' is called the *implication connective*, such statements are sometimes called *implication statements*.

Everything stated below regarding conditional statements applies to each of the forms described above.

B. How to negate a conditional statement

Let us examine two statements:

p: It is raining in Wonderland.
q: The White Rabbit is wet.

We assume that $p \rightarrow q$ holds true. In other words, the statement "If it is raining in Wonderland, then the White Rabbit is wet" is a true statement; that is, its content is true. Clearly its negation, therefore, is false.

But what is the negation of this statement?

Since we assumed that the conditional statement is true, obviously it could not be that it is raining in Wonderland and the White Rabbit is **not** wet. The negation of the conditional statement is therefore the false statement "It is raining in Wonderland **and** the White Rabbit is **not** wet." Note how the connective AND made its way into the conditional statement.

In general, negation of a conditional statement of the form "IF p, THEN q" is "p AND NOT q" or, in symbolic notation: $p \wedge \sim q$.

C. What can be inferred from a conditional statement

In this section we present four different methods for expressing a conditional statement using equivalent statements; that is, statements which all follow from the conditional statement, and the conditional statement follows from each of them.

1. We found that the negation of the conditional statement $p \rightarrow q$ is $p \wedge \sim q$. It follows from this that the conditional statement $p \rightarrow q$ is equivalent to the statement: $(p \wedge q)$ OR $(q \wedge \sim p)$ OR $(\sim q \wedge \sim p)$ (because there are only four possibilities). This may be written as:

$$p \rightarrow q \equiv (p \wedge q) \vee (\sim p \wedge q) \vee (\sim p \wedge \sim q)$$

Thus, for example, from the statement "It is raining in Wonderland, then the White Rabbit is wet," the following statement can be inferred:

"It is raining in Wonderland and the White Rabbit is wet, or it is not raining in Wonderland and the White Rabbit is wet, or it is not raining in Wonderland and the White Rabbit is not wet." The only case that cannot hold is that "It is raining in Wonderland and the White Rabbit has remained dry." Note that the conditional statement says nothing about the Rabbit's condition in the event that it is not raining.

2. Since the negation of the conditional statement $p \to q$ is: $p \wedge \sim q$, it follows that $p \to q$ is **equivalent** to the negation of its negation; that is, it is equivalent to $\sim(p \wedge \sim q)$:

$$p \to q \equiv \sim(p \wedge \sim q)$$

So, for example, from the statement "If it is raining in Wonderland, then the White Rabbit is wet", it can be inferred that "It cannot be the case that it is raining in Wonderland and the White Rabbit is not wet."

The reverse is true as well — from the statement "It cannot be the case that it is raining in Wonderland and the White Rabbit is not wet," it can be inferred that "If it is raining in Wonderland, then the White Rabbit is wet."

3. In the discussion of De Morgan's Laws in Chapter 4, we found the equivalence:

$$\sim(p \wedge \sim q) \equiv \sim p \vee q$$

In Section C2 above we found that:

$$p \to q \equiv \sim(p \wedge \sim q)$$

It follows from this (due to the transitive nature of equivalence) that:

$$p \to q \equiv \sim p \vee q$$

In other words, the conditional statement $p \to q$ and the statement $\sim p \vee q$ are **equivalent** statements as well. So, for example, from the statement "If it is raining in Wonderland, then the White Rabbit is wet," it can be inferred that "It is not raining in Wonderland or the White Rabbit is wet." The reverse is true as well — from the statement "It is not raining in Wonderland or the White Rabbit is wet" it can be inferred that "If it is raining in Wonderland, then the White Rabbit is wet."

4. We stated that if it is raining in Wonderland, then the White Rabbit is wet. Could it be that the White Rabbit is not wet, and it is raining in Wonderland? — No. We saw in Section B above that this could not be the case. In other words, if the White Rabbit is not wet, then, without a doubt, it is not raining in Wonderland. This last statement is a conditional statement as well. It is different from the original statement, but expresses the identical content. The difference between them is that the antecedent and the consequent have reversed roles and in parallel have taking on a negation (that is, their truth values have flipped). In general, the new statement $\sim q \to \sim p$ is **equivalent** to the original $p \to q$. This may be expressed alternatively as:

$$p \to q \equiv \sim q \to \sim p$$

These are referred to as **contrapositives** of one another.

D. What cannot be inferred from a conditional statement

Once again let us examine the statement "If it is raining in Wonderland, then the White Rabbit is wet." We saw earlier that this statement does not indicate anything about the White Rabbit when it is not raining in Wonderland. Note in particular that the conditional statement "If it is **not** raining in Wonderland, then the White Rabbit is **not** wet" **does not** follow from the given statement. The White Rabbit could be wet even if it is not raining in Wonderland (it may have gotten wet from a water sprinkler, for example). In general, it is an error (and a common one) to infer the **inverse**, that is, $\sim p \to \sim q$, from the statement $p \to \sim q$.

Similarly, it is an error (again, a common one) to infer the **converse**, $q \to p$, from the statement $p \to q$. The statement "If it is raining in Wonderland, then the White Rabbit is wet" does not indicate that "If the White Rabbit is wet, then it is raining in Wonderland." As we indicated, the White Rabbit could have gotten wet from a water sprinkler.

The tendency to infer erroneously from the statement $p \to q$ one of the statements: $\sim p \to \sim q$ or $q \to p$, apparently stems from a sense of verbal symmetry or from imprecise use of "IF ..., THEN ..." in daily language. An example of this is the use of "IF ..., THEN ..." as a promise or a threat: A child who is promised "If you finish your dinner you will get dessert," understands this as a threat — if he does not finish his dinner, he will not get dessert. This interpretation encourages him to finish his dinner, but in truth, the threat does not follow from the promise made to him, although it is true that was the intent. According to the promise made, he may get dessert even if he does not finish his dinner. A child who is told "If it rains I will pick you up from school" may justifiably ask: "And what if it doesn't rain?" The answer does not follow from the promise, although it may be assumed that what is meant is "ONLY if it rains will I pick you up from school; if it doesn't rain, I will not."

E. The truth table for the implication connective

Let us construct a truth table for the two statements p and q, and for the compound statement $p \rightarrow q$, obtained by using the logical connective 'IF ..., THEN ...'. Recall that there are four combinations of truth values for the statements p and q:

p	q
T	T
T	F
F	T
F	F

In Section C above we found that the conditional statement $p \rightarrow q$ is equivalent to each of the following three compound statements:

$$(p \wedge q) \vee (\sim p \wedge q) \vee (\sim p \wedge \sim q),$$

$$\sim(p \wedge \sim q),$$

$$\sim p \vee q$$

We construct the truth table for the latter equivalent statement, using our knowledge of the NOT and AND connectives from Chapters 1 and 2.

p	q	$\sim p$	$\sim p \vee q$
T	T	F	T
T	F	F	F
F	T	T	T
F	F	T	T

But, as indicated:

$$p \rightarrow q \equiv \sim p \vee q$$

and therefore the truth table for the implication connective is:

p	q	$p \rightarrow q$
T	T	T
T	F	F
F	T	T
F	F	T

Remark:

• Students may be called upon to prepare a truth table for one of the first two equivalent statements. This should be done based on the knowledge they acquired on the connectives OR, AND and NOT in Chapters 1–3. The results, of course, are identical.

F. Rules of inference

The rules of inference provide a template for determining that a particular conclusion necessarily follows from a given set of data. There are two basic rules of inference:

Inference Rule #1: Given the conditional statement "IF p, THEN q" that is, given that $p \rightarrow q$ is a true statement, and also given p that is, that the antecedent of the conditional statement is also a true statement, it follows that the consequent statement q is necessarily true as well.

For example:

- Given: 1. If it is raining in Wonderland, then the White Rabbit is wet.
           2. It is raining in Wonderland.

   Conclusion: The White Rabbit is wet.

This rule of inference is called **modus ponendo ponens**, or **modus ponens** for short. This is the most basic rule of inference.

It is written in statement notation as:

$$p \rightarrow q$$
$$p$$
$$\therefore \qquad q$$

Note: This rule is also known as **The Law of Detachment**.

Inference Rule #2: Given the conditional statement "IF p, THEN q" that is, given that $p \rightarrow q$ is a true statement, and also given $\sim q$ that is, that the consequent of the conditional statement is false, then it follows necessarily that $\sim p$. That is, the antecedent statement p is false as well.

For example:

- Given: 1. If it is raining in Wonderland, then the White Rabbit is wet.
           2. The White Rabbit is not wet.

   Conclusion: It is not raining in Wonderland.

This inference rule is known as **modus tollendo tollens**, or **modus tollens**, for short.

It is written in statement notation as:

$$p \rightarrow q$$
$$\sim q$$
$$\overline{}$$
$$\therefore \ \sim p$$

Notes:

- We saw in Section 4C that the conditional statement "IF p, THEN q" is equivalent to its contrapositive "IF $\sim q$, THEN $\sim p$". Application of modus tollens, therefore, is equivalent to application of modus ponens. It is written in statement notation as:

$$\sim q \rightarrow \sim p$$
$$\sim q$$
$$\overline{}$$
$$\therefore \qquad \sim p$$

- As mentioned above, it is important to distinguish between a causal relationship that may be expressed by the content of the conditional statement, and its structure. Thus, it should be noted that the rules of inference determine an outcome based on the structure of the given statements and not on their content.

For example:

Whoever assumes that the statements given below are true must draw the conclusion that the White Rabbit will buy a new watch — even though the conditional statement has a causal relationship that is unclear to us. The statements are:

- If ears of corn grow in Wonderland, then the White Rabbit will buy a new watch
- Ears of corn grow in Wonderland

G. Logical fallacies in drawing conclusions

As mentioned above, there is a natural tendency to infer from the truth of the conditional statement "IF p, THEN q" that it is also true that "IF NOT p, THEN NOT q." This fallacy may be expressed as:

$$p \rightarrow q$$
$$\sim p$$
$$\overline{}$$
$$\therefore \qquad \sim q$$

For example, assuming that the statement "If it is raining in Wonderland, then the White Rabbit is wet" is true, and someone in Wonderland is looking out of the window and sees that it is not raining, does this mean that the White Rabbit is not wet? This question cannot be answered conclusively. The given information does not suffice to answer the question. The White Rabbit may be wet and it may not be wet. What we do know is what happens to the White Rabbit when it **is** raining in Wonderland. The conditional statement does not tell us anything about what happens to the White Rabbit when it is **not** raining in Wonderland.

$$p \rightarrow q$$

$$\sim p$$

There is insufficient information to draw
conclusions one way or the other.

Such a logical fallacy is known as **denying the antecedent**.

An additional kind of logical fallacy is **accepting the consequent**.

Let us assume that we met the White Rabbit and saw that it is wet. Does this mean that it is raining in Wonderland?

Clearly it cannot be determined conclusively whether the White Rabbit is wet from the rain, from a water sprinkler, or that it just came out of the shower. Therefore no conclusion can be drawn regarding rain in Wonderland.

This fallacy may be written as follows:

$$p \rightarrow q$$

$$q$$

$$\therefore p$$

The truth is that

$$p \rightarrow q$$

$$q$$

There is insufficient information to
draw any conclusions.

H.   Bi-directional conditional statements

p, q are two statements. Two conditional statements may be constructed from them: $p \rightarrow q$ and $q \rightarrow p$. We say that each conditional statement is the **converse** of the other. We saw above that one of the conditional statements may be true, while the other doesn't necessarily follow from it; that is, perhaps it is true, and perhaps it is false. In other words, all possibilities are open — both could be false statements, both could be true statements, or one could be true and the other false. In the case that both conditional statements are true, that is, $p \rightarrow q$ and $q \rightarrow p$ both hold, then we also say *p* if and only if *q*, or *q* if and only if *p*. This is notated in short as:*p* iff *q*, and is denoted by $p \leftrightarrow q$.

Examples:

- *p*: A given quadrilateral is a parallelogram.

  q: The diagonals of this quadrilateral bisect each other.

  The resultant statements are as follows:

    If a quadrilateral is a parallelogram (*p*), then its diagonals bisect each other (*q*).
    If the diagonals of a quadrilateral bisect each other (*q*), then the quadrilateral is parallelogram (*p*).

  Since both of the two conditional statements are true, we can state that a quadrilateral is a parallelogram if and only if its diagonals bisect each other, or that diagonals in a quadrilateral bisect each other if and only if the quadrilateral is a parallelogram.

- Two triangles are congruent if and only if the three sides of one are congruent to the corresponding three sides of the other respectively.

- The quadratic equation $ax^2 + bx + c = 0$ in which $a \neq 0$, $a, b, c \in R$, has two distinct solutions if and only if the discriminant $b^2 - 4ac > 0$.

  Notes:

  - Decomposition of the conditional statement into its respective components must be performed carefully. For example, in the last case presented above, the conditions that apply to *a, b, c,* were excluded from the second half of the bi-directional conditional statement due to linguistic considerations. Upon decomposition, these need to be restated.

- The facts presented in Section C above with regard to statements equivalent to conditional statements can now be written as follows:

$$(p \rightarrow q) \leftrightarrow [(p \wedge q) \vee (\sim p \wedge q) \vee (\sim p \wedge \sim q)]$$

$$(p \rightarrow q) \leftrightarrow [\sim(p \wedge \sim q)]$$

$$(p \rightarrow q) \leftrightarrow [\sim p \vee q]$$

$$(p \rightarrow q) \leftrightarrow [\sim q \rightarrow \sim p]$$

- Students may be called upon to try phrasing De Morgan's laws (Chapter 4) using bi-directional conditional statements.

- As indicated, in logic — the language of mathematics — the implication connective is denoted by convention using an arrow $\rightarrow$, and the bi-directional conditional statement is denoted using a double arrow $\leftrightarrow$. On the other hand, in mathematics these are conventionally indicated with the symbols $\Rightarrow$, $\Leftrightarrow$ rather than $\rightarrow$, $\leftrightarrow$, respectively.

Examples:

- $4|a \Rightarrow 2|a$ for each a that is a natural number;

- $(\triangle ABC \cong \triangle DEF) \Leftrightarrow [(AB = DE) \wedge (BC = EF) \wedge (CA = FD)]$.

In Chapter 9 we discuss the meaning of "if and only if" once again, in the context of necessary and sufficient conditions.

8.8. Review of Answers to Worksheet 8b

After students have become familiar with the logical background, it is advised to ask them to go back and review their answers to Worksheet 8b and correct them if necessary.

8.9. Worksheet 8c: The Logical Connective 'IF..., THEN...' – Drawing Conclusions Using Rules of Inference

Worksheet 8c addresses inference from statements containing the logical connective 'IF ..., THEN ...'.

Remarks:

- This worksheet may be completed as individual practice, in pairs or in small groups, with group discussion encouraged.
- Upon completion of the worksheet, it is recommended that the students present their answers to the class to initiate a whole class discussion of the solutions.

Worksheet 8c and Proposed Solutions

Each part of the table below contains two statements followed by a conclusion. Rewrite the given statements and the proposed conclusion using statement notation. Then examine the proposed conclusion in light of the given statements, and determine whether the conclusion does in fact follow from them.

Part	Given	The Conclusion
1	A. If you hold a red-hot poker too long, then it will burn you. B. I got a burn on my finger.	I held a red-hot poker too long.

Let us designate the statements as follows:

    p: (People) hold red-hot pokers for too long.

    q: Red-hot pokers burn.

We now indicate the given information in statement notation.

$$p \rightarrow q$$
$$q$$
$$\overline{}$$
$$\therefore p$$

We construct a truth table corresponding to the logical connective 'IF ..., THEN ...'. It is given that $p \rightarrow q$ is TRUE. This bit of information corresponds to the first, third and fourth rows of the truth table for the implication connective.

p	q	$p \rightarrow q$
T	T	T
T	F	F
F	T	T
F	F	T

It is also given that q is a TRUE statement. This narrows the possibilities to the first and third rows.

From these two rows we see that statement p could be TRUE or FALSE; therefore it cannot be determined whether or not the proposed conclusion presented in the table ("I held a red-hot poker too long") is TRUE.

p	q	$p \rightarrow q$
T	T	T
T	F	F
F	T	T
F	F	T

Or, phrased in words: Although it is possible that I got a burn because I held a red-hot poker too long, it is also possible that the burn was caused by something else. (A common logical fallacy would be that the conclusion is TRUE.)

Part	Given	The Conclusion
2	A. If you cut your finger *very* deeply with a knife, then it bleeds. B. My finger is not bleeding.	I did not cut a *very* deep cut on my finger with a knife.

Let us designate the statements as follows:

p: (People) cut a very deep cut on their finger with a knife.

q: (Someone's) finger is bleeding.

We now indicate the given information in statement notation.

$$p \rightarrow q$$
$$\sim q$$
$$\overline{}$$
$$\therefore \ \sim p$$

We construct a truth table corresponding to the logical connective 'IF ..., THEN ...'.

It is given that the compound statement $p \rightarrow q$ is TRUE. For this reason, just as in the case in the first question,

in accordance with the truth table for this connective, the first, third and fourth rows match the given information.

It is also given that q is FALSE. Thus only the fourth row matches the given information. From this row we can see that statement p is FALSE; therefore, the proposed conclusion in the table is TRUE.

p	q	$p \rightarrow q$
T	T	T
T	F	F
F	T	T
F	F	T

Or, phrased in words: According to the first bit of information, it cannot happen that there is a very deep cut in the finger and the finger is not bleeding. Therefore, the second piece of information, which establishes that the finger is not bleeding, invalidates the possibility that the finger was cut.

Part	Given	The Conclusion
3	A. If you drink from a bottle marked "poison", then it will disagree with you. B. Alice did not drink from a bottle marked "poison".	Nothing disagreed with Alice.

Let us designate the statements as follows:

p: (People) drink from a bottle marked "poison".

q: Something disagreed with (people).

We now write the given information in statement notation:

$$p \rightarrow q$$

$$\sim p$$

$$\therefore \quad \sim q$$

We construct a truth table corresponding to the logical connective 'IF ..., THEN ...'.

It is given that the compound statement $p \rightarrow q$ is TRUE. For this reason, just as in the case in the first question, in accordance with the truth table for this connective, the first, third and fourth rows of the table match the given information.
It is also given that p is FALSE. Only the third and fourth rows match this information.

p	q	$p \rightarrow q$
T	T	T
T	F	F
F	T	T
F	F	T

From these two rows we see that statement q can be either TRUE or FALSE; therefore it cannot be determined whether or not the proposed conclusion presented in the table ("Nothing disagreed with Alice") is TRUE.

Or, phrased in words: All we know is that nothing disagreed with Alice as a result of drinking from the bottle of poison, but this does not guarantee that there was not something that disagreed with her as a result of something else.

Part	Given	The Conclusion
4	A. If the bottle is not marked "poison", then the liquid inside tastes like custard. B. The bottle is not marked "poison".	The liquid in the bottle tastes like custard.

Let us designate the statements as follows:

  p: The bottle is not marked "poison".

  q: The liquid in the bottle tastes like custard.

We now indicate the given information in statement notation.

$$p \to q$$

$$p$$

$$\therefore \quad q$$

We construct a truth table corresponding to the logical connective 'IF ..., THEN ...'.

It is given that the conditional statement $p \to q$ is TRUE. For this reason, just as in the case in the first question, in accordance with the truth table for this connective, the first, third and fourth rows of the table match the given information.

It is also given that p is a TRUE statement. Only the first and second rows match this information. Therefore only the first row matches both bits of information.

p	q	$p \to q$
T	T	T
T	F	F
F	T	T
F	F	T

From this row we obtain that statement q must be TRUE; therefore the proposed conclusion is TRUE. Or, phrased in words: The conclusion is true by applying the first rule of inference (modus ponens).

8.10. Worksheet 8d: The Logical Connective 'IF…, THEN…' — The Four-Card Problem

The problem included in this worksheet is known as "the four-card problem" or the "Wason selection task." The problem (unrelated to Alice, of course), was first presented in 1972 by two psychologists: Peter Wason and Philip Johnson-Laird.[1]

Worksheet 8d and Proposed Solutions

1. Alice and the White Rabbit like to ask each other riddles. One day Alice laid four cards on the table, with only one side of them visible.

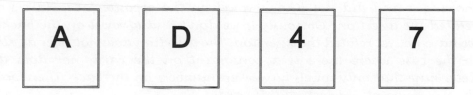

Alice told the White Rabbit that on one face of each card is a letter, and on the other face is a number.

Then Alice posed the following question to the White Rabbit:

"Which card or cards must be turned over in order to prove or disprove the following claim:

For each card, if a card shows a vowel on one face, then its opposite face has an even number."

What would you answer if you were the White Rabbit?

In 1972, when the two psychologists Wason and Johnson-Laird presented the problem to 128 students, 59 of them (46%) claimed that the cards showing A and 4 must be turned over. Forty-two students (33%) claimed that the card showing A must be turned over. Only 5% of the students gave the right answer.

---

[1] Additional details may be found in the book:

Wason, P. C. & Johnson-Laird, P. N. (1972). *Psychology of Reasoning: Structure and Content.* Harvard University Press.

The most common reasoning given for choosing the card showing A was: If on the opposite side of the card is an even number, then the assertion is validated at least partially. On the other hand, if on the opposite face is an odd number, then the assertion has been refuted.

The reasoning behind this explanation is: A is a vowel. Let us assume that the assertion is true. If so, on the opposite side of the card there must be an even number. If we flip the card and find an odd number, then the assertion is invalid. This thinking leads to proof by negation, and it is in fact necessary in this case to turn this card over.

The most common explanation for selecting the card showing 4 was: We will check whether there is a vowel on the other side. This answer comes from a logical fallacy. Let us assume that we turned over the card showing 4 and did not find a vowel on the opposite face. Have we refuted the assertion? Obviously if we don't find a vowel on the back, we have not yet refuted the assertion. The assertion does not refer at all to the case where there is a consonant on the card, nor does it determine that only vowels have even numbers on the back. Therefore there is no point in selecting this card.

The other card that must be turned over in order to examine the truth of the assertion is the card showing the number 7. For the assertion to be true, a vowel may not appear on the back of the card, since 7 is an odd number. If a vowel appears on the back of the card, then the assertion has been refuted. Here too, the proof is by negation.

Would there be a point in turning over the card showing D? — Let us assume that the card showing D was turned over, and on the back was an even number. Has the assertion been proven by this? Let us assume that on the other side was an odd number. Has the assertion been refuted by this? The answer to both of these questions is no, since the assertion does not relate to consonants. Therefore there is no point in turning this card over.

To summarize — the cards that must be turned over are the ones labeled A and 7. It is possible that the assertion will be refuted after turning over only one of them. But if the first card supports the assertion, then the second card must still be turned over, in order to see whether it too supports the assertion.

Remark:

- Wason and Johnson-Laird presented another version of the problem, in which they retained the same assertion, but presented only two cards — on one there was a 1 and on the other a 2. You may wish to present this problem to students after the discussion of Worksheet 8d.

8.11. Worksheet 8e: The Logical Connective 'IF..., THEN...' — Negation of the Conditional Statement

Worksheet 8e addresses the negation of the conditional statement.[2]

Remarks:

- This worksheet may be completed as individual practice, in pairs or in small groups, with group discussion encouraged.
- Upon completion of the worksheet, students may present their answers to the class; a whole class discussion of these solutions is recommended.

---

[2] This worksheet was inspired by the following sources:

Hadar, N. (1975). *Children's Conditional Reasoning.* Ph.D.dissertation. University of California, Berkeley. ERIC accession number ED118359, RIE June 1976.

Hadar, N. (1977). An intuitive approach to the logic of implication. *Educational Studies in Mathematics, 8,* 413–438.

Worksheet 8e and Proposed Solutions

A. In each picture below there is a White Rabbit and a TV set. Examine each picture carefully and read the sentence that appears below them. Then, circle the numbers of the pictures that contradict the content of the statement. Explain your answer.

1

2

3

4

5

6

If the TV is on, then the White Rabbit is not playing the trumpet

Picture 3 contradicts the content of the statement. The conditional statement is: If the TV is on, then the White Rabbit is not playing the trumpet. Each picture that presents a TV set that is on alongside a Rabbit that is playing the trumpet contradicts this statement. This can be stated formally as follows:

p: The TV set is on.

q: The White Rabbit is not playing the trumpet.

The given conditional statement: $p \rightarrow q$ is equivalent to $\sim(p \wedge \sim q)$, and therefore its contradiction is $p \wedge \sim q$. Note that the cases $\sim p \wedge \sim q$, $\sim p \wedge q$ do not contradict the conditional statement; that is, the pictures in which the TV is off do not contradict the conditional statement, since the conditional statement does not provide any information about the behavior of the Rabbit when the TV is off. Alongside a TV that is turned off, there could be two possible cases — either the Rabbit is playing the trumpet, or it is not.

B.   In each picture below there is a White Rabbit and a TV set. Examine each picture carefully and read the sentence that appears below them. Then, circle the numbers of the pictures that **contradict** the content of the statement. Explain your answer.

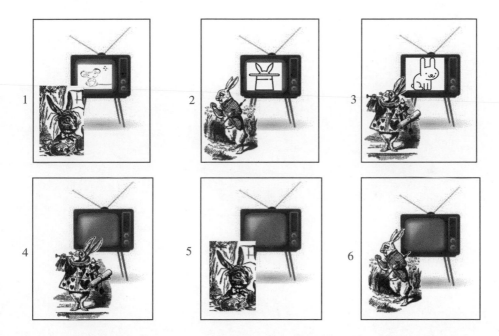

If the White Rabbit is not playing the trumpet, then the TV is on.

Pictures 5 and 6 contradict the content of the statement.

The conditional statement here is: If the White Rabbit is not playing the trumpet, then the TV is on. Note that this conditional statement is the converse of the previous one (see Section 8.7h above). Every picture that presents the White Rabbit not playing the trumpet alongside a TV that is not on, contradicts this statement. This can be stated formally as follows:

p: The White Rabbit is not playing the trumpet.

q: The TV set is on.

The conditional statement: $p \to q$ is equivalent to $\sim(p \wedge \sim q)$; therefore its contradiction is $p \wedge \sim q$. Note that the cases $\sim p \wedge \sim q$, $\sim p \wedge q$ do not contradict the conditional statement; that is, the pictures in which the Rabbit is playing the trumpet do not contradict the conditional statement, since the conditional statement provides no information about the state of the TV when the White Rabbit is playing the trumpet.

8.12. Worksheets 8f and 8g: Summary Exercise for Conditional Statements

The summary exercise is composed of two worksheets: 8f and 8g.

Remarks:

- Worksheet 8f is to be completed as individual practice by students. In Worksheet 8g (which appears in the students' workbook only) students will compare their current answers with the answers they gave at the start of the chapter (Worksheet 8a).
- Upon completion of the summary exercise, it is recommended that a whole class discussion be held in order to consider the changes that have taken place in students' understanding and perceptions through the course of this chapter, as well as to identify the particular difficulties encountered with the subject matter.

Worksheet 8f and Proposed Solutions

Answer the questions in this worksheet in their entirety, providing as much detail as possible. If necessary, you may indicate: "I did not understand the question, therefore I have not answered it." Make sure not to go back to Worksheet 4a before you complete your work on this worksheet.

Below is a table with four items. Each item contains two statements followed by a proposed conclusion.

For each proposed solution, select one of the following three options, and explain your choice:

A.   It follows from the statements that the conclusion is true.

B.   It does not necessarily follow from the statements that the conclusion is true.

C.   It follows from the statements that the conclusion is false.

	The given statements	Proposed conclusions	It follows from the statements that the conclusion is true	It does not necessarily follow from the statements that the conclusion is true	It follows from the statements that the conclusion is false
A	1. If the White Rabbit comes to the party, then Alice will be happy. 2. The White Rabbit did not come to the party.	Alice is happy.		*It cannot be determined whether or not Alice is happy, since the conditional statement provides information about Alice only in the case in which the Rabbit comes to the party.*	

(Continued)

(*Continued*)

	The given statements	Proposed conclusions	It follows from the statements that the conclusion is true	It does not necessarily follow from the statements that the conclusion is true	It follows from the statements that the conclusion is false
B	1. If the White Rabbit comes to the party, then Alice will be happy. 2. Alice is not happy.	The White Rabbit came to the party.			*The given conditional statement assures that when the proposed conclusion is true, Alice is happy. But it is also given that Alice is not happy. Therefore it could not be that the proposed conclusion is true.*
C	1. If the White Rabbit comes to the party, then Alice will be happy. 2. The White Rabbit came to the party.	Alice is happy.	*The given conditional statement assures that Alice will be happy if the White Rabbit comes to the party. When the White Rabbit does in fact come to the party, Alice is happy.*		

(*Continued*)

(Continued)

	The given statements	Proposed conclusions	It follows from the statements that the conclusion is true	It does not necessarily follow from the statements that the conclusion is true	It follows from the statements that the conclusion is false
D	1. If the White Rabbit comes to the party, then Alice will be happy.

2. Alice is happy. | The White Rabbit came to the party. | | *It cannot be determined whether the White Rabbit came to the party or not. Perhaps Alice is happy for other reasons. The conditional statement provides information about one of the reasons for Alice's happiness, but does not establish this as the only reason.* | |

Necessary and Sufficient Conditions

A NECESSARY condition
for Alice to be able to go
into the garden is for her
to be little.
A SUFFICIENT condition
for the Rabbit to know
the time is for it to have a
watch in its waistcoat-
pocket.

Chapter 9

Necessary and Sufficient Conditions

A NECESSARY condition
for Alice to be able to go
into the garden is for her
to be tall.

A SUFFICIENT condition
for the Rabbit to know
the time is for it to have a
watch in its waistcoat-
pocket.

9.1. Worksheet 9a: Opening Exercise. Lead-in to Necessary and Sufficient Condition

Instruction of this chapter begins with Worksheet 9a Opening Exercise (here forth abbr. WS9a), to be completed as individual practice by each student. Upon completion of the task, notify the students that they will return to this worksheet at the end of the chapter (Worksheet 9e Summary Exercise); then they will be able to compare their initial understanding with that which they acquired during the course of the chapter, while reflecting upon the changes in their understanding along with the points where further clarification is still required (Worksheet 9f, which appears only in the Student's Workbook). For this reason it is recommended that discussion of this worksheet will be held off until the end of the chapter. In some cases the teacher may prefer to collect WS9a and give it back only upon completion of WS9e when students are ready to work on WS9f.

Because this lead-in worksheet is repeated again as a summary exercise, the worksheet (together with proposed solutions) appears only once — at the end of this chapter.

9.2. A Look Inside the Text

Read the paragraphs from page 10 of *Alice's Adventures in Wonderland* that appear in boldface below. This excerpt will serve as the basis for discussion for the topic 'Necessary and Sufficient Condition.'

However, this bottle was *not* marked "poison," so Alice ventured to taste it, and finding it very nice (it had, in fact, a sort of mixed flavour of cherry-tart, custard, pineapple, roast turkey, coffee, and hot buttered toast,) she very soon finished it off.

"What a curious feeling!" said Alice. "I must be shutting up like a telescope."

And so it was indeed: she was now only ten inches high, and her face brightened up at the thought that she was now the right size for going through that little door into that lovely garden. First, however, she waited for a few minutes to see if she was going to shrink any further: she felt a little nervous about this: "for it might end, you know," said Alice to herself, "in my going out altogether, like a candle. I wonder what I should be like then?" And she tried to fancy what the

flame of a candle looks like after the candle is blown out, for she could not remember ever having seen such a thing.

After a while, finding that nothing more happened, she decided on going into the garden at once; but, alas for poor Alice! when she got to the door, she found she had forgotten the little golden key, and when she went back to the table for it, she found she could not possibly reach it: she could see it quite plainly through the glass, and she tried her best to climb up one of the legs of the table, but it was too slippery; and when she had tired herself out with trying, the poor little thing sat down and cried.

9.3. Worksheet 9b: Necessary and Sufficient Condition

The central topic of this chapter — necessary and sufficient condition — kicks off with Worksheet 9b, which begins with the excerpt shown above (Section 9.2).

Remarks:

- Worksheet 9b is to be completed in pairs or in small groups to allow for consultation.
- It is suggested that upon completion of the worksheet, students present their answers to the class, in order to initiate a whole class discussion of the solutions.
- At this point it is advised to avoid any judgment of students' answers; points of disagreement and misperceptions, however, should be noted.
- After presentation of the logical and mathematical background by the teacher (Section 9.4 below), students' answers should be revisited and discussed once again.

Worksheet 9b and Proposed Solutions

1. According to the excerpt above, does it suffice for Alice's size to match that of the door in order for her to be able to enter the garden?

    No. It does not suffice for Alice's size to match that of the door in order for her to be able to enter the garden. As Alice discovered, she also needs the key in order to enter the garden. Being the appropriate size is necessary in order for Alice to pass through the door. But it is not sufficient.

2. According to the excerpt above, does it suffice for Alice to have the key to the little door in order for her to enter the garden?

    No. It does not suffice for Alice to have the key to the little door in order for her to be able to enter the garden. Although she needs the key in order to be able to enter the garden, if her size does not change to match that of the door, she will not be able to pass through it even if the key is in her hand. The key is necessary in order for Alice to pass through the door, but it is not sufficient.

3. What suffices in order for Alice to be able to enter the garden through the little door?

    In order for Alice to be able to enter the garden through the little door, it suffices for both conditions to exist simultaneously: her size must match that of the little door, and she must have the key for unlocking the little door. Existence of both of these suffices to guarantee that Alice could enter the garden through the little door.

4. Are you familiar with a situation in mathematics similar to that which Alice encountered with respect to the two conditions for entering the garden through the little door?

    Divisibility by 2 is necessary for a number to be divisible by 6. But it is not sufficient (there are other numbers such as 10, for example, that are divisible by 2 but are not divisible by 6). Divisibility by 3 is also necessary for a number to be divisible by 6, but it is not sufficient. (Why?) On the other hand, divisibility by 2 and by 3 is sufficient for a number to be divisible by 6.

9.4. Logical and Mathematical Background

Upon completion of Worksheet 9b, and after discussion of students' answers, the following concepts and topics are to be presented:

    A. A necessary condition
    B. A sufficient condition
    C. A necessary condition that is not sufficient, a sufficient condition that is not necessary, and a necessary and sufficient condition
    D. The relationship between a necessary condition and a sufficient condition
    E. The relationship between a necessary condition, a sufficient condition, and a conditional statement
    F. The relationship between a necessary condition, a sufficient condition, and statements with the quantifiers ALL or ONLY
    G. The meaning of "definition" and its relationship to necessary condition, sufficient condition, and to necessary and sufficient condition

A. A necessary condition

p and q are statements. When we say that p is a **necessary condition** for q, we mean that q can be TRUE **only** if p is TRUE. In other words, if p is **not** TRUE, then it is certain that q is not TRUE as well. The only conclusion that may be inferred is that if q is TRUE, p is TRUE with certainty. But even if p is TRUE, **it cannot be inferred** that q is TRUE as well.

For example:

- Given two statements:

  p: There are clouds in the sky now.
  q: It is raining now.

Is p a necessary condition for q? The answer is 'Yes' — p is a necessary condition for q. In order for it to rain, it is necessary for there to be clouds. That is, clearly if it is raining now (q is TRUE), then there are necessarily clouds in the sky (p is TRUE). If it is **FALSE** that there are clouds in the sky now (p is FALSE), then it is **FALSE** with certainty that it is raining now (q is FALSE). But it is also possible that there are clouds in the sky now (p is TRUE), yet it is not raining now (q is

FALSE). Therefore we say that the clouds are a necessary condition for rain, or, **only if** there are clouds is it raining.

- Given two statements:

  s: A given polygon has four sides.
  t: A given polygon is a square.

Is *s* a necessary condition for *t*? The answer is 'Yes' — *s* is a necessary condition for *t*. In order to a polygon to be a square, it necessarily has four sides. That is, if the given polygon is a square (*t* is TRUE), then the polygon necessarily has four sides (*s* is TRUE). In other words, if the number of sides of the given polygon is **not** four (*s* is FALSE), then it cannot be a square (*t* is FALSE). Yet, even if the given polygon **does have** four sides, it is not necessarily a square. It could, for example, be a trapezoid. We therefore say that having four sides is a necessary condition for a polygon to be a square, or, a polygon is a square **only if** it has four sides.

- Given two statements:

  m: The units digit of a number is 2.
  n: The number is even.

Is *m* a necessary condition for *n*? The answer to this question is "No". As indicated, the meaning of "*m* is a necessary condition for *n*" is that if *m* is FALSE, then *n* is FALSE with certainty. In this case, it is possible that the units digit of a given number is not 2 (for example, 6), but the number is still even (for example, 36).

B. A sufficient condition

p and *q* are statements. When we say that *p* is a **sufficient condition** for *q*, we mean that if *p* is TRUE, then *q* is TRUE. In other words, if *q* is **not** TRUE, then *p* is **not** TRUE either. Yet *q* **could** hold TRUE even if *p* is **not** TRUE.

For example:

- Given two statements:

  p: I have a dog.
  q: I have a pet.

Is *p* a sufficient condition for *q*? The answer is 'Yes' — *p* is a sufficient condition for *q*. If I have a dog (*p* is TRUE), then clearly I have a pet (*q* is TRUE). But it is also possible that I have a pet (*q* is TRUE) yet the pet is not a dog (*p* is FALSE), but rather it is a cat, for example. Still, it is clear that if I **do not** have a pet (*q* is FALSE), then with certainty I do not have a dog (*p* is FALSE).

- Given two statements:

  r: A given quadrilateral is a square.
  s: The quadrilateral has four right angles.

Is *r* a sufficient condition for *s*? The answer is 'Yes' — *r* is a sufficient condition for *s*, since if a quadrilateral is a square (*r* is TRUE), it is sufficient to guarantee that the quadrilateral has four right angles (*s* is TRUE). But it is also possible that the quadrilateral has four right angles (*s* is TRUE), but the quadrilateral is not a square (*r* is FALSE), but an oblong rectangle.

- Given two statements:

  g: A number *a* is divisible by five.
  h: The units digit of the number *a* is 5.

Is *g* a sufficient condition for *h*? The answer is 'No' — *g* is not a sufficient condition for *h*, since if a number is divisible by five (*g* is TRUE), it is possible that the units digit is not five (*h* is FALSE), but rather zero. On the other hand, *h* is a sufficient condition for *g*.

C. A necessary condition that is not sufficient, a sufficient condition that is not necessary, and a necessary and sufficient condition

Necessary conditions and sufficient conditions are independent of one another. A necessary condition could be not sufficient, and a sufficient condition could be not necessary.

For example:

- Divisibility by 2 is a necessary condition for divisibility by 6, but is not a sufficient condition for it (there are numbers that are divisible by 2 but are not divisible by 6; 8 is one such example).

- Divisibility by 12 is a sufficient condition for divisibility by 6, but it is not a necessary condition for it (the number 18, for example, is divisible by 6 but is not divisible by 12).
- On the other hand, divisibility by both 2 and 3 is a necessary condition and a sufficient condition for divisibility by 6.

There are many cases where a necessary condition is also a sufficient condition, or conversely — cases where a sufficient condition is also a necessary condition. In these cases we say that the condition is necessary and sufficient. When a condition is both necessary and sufficient, we say that this condition can serve as a definition (see Section G below).

D. The relationship between a necessary condition and a sufficient condition

We begin with the examples already introduced above:

- Divisibility by 2 is a necessary condition for divisibility by 6. This is equivalent to the assertion: Divisibility by 6 is a sufficient condition for divisibility by 2.
- Divisibility by 12 is a sufficient condition for divisibility by 6. This is equivalent to the assertion: Divisibility by 6 is a necessary condition for divisibility by 12.

In general, if p is a **necessary** condition for q, then q is a **sufficient** condition for p, and conversely — if q is a **sufficient** condition for p, then p is a **necessary** condition for q.

This can be described using a Venn diagram:

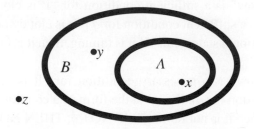

For example, A is the set of squares; B is the set of quadrilaterals. Since $A \subset B$, the existence of an element x in A is a sufficient condition for its existence in B.

For the same reason, existence of an element in B is a necessary condition for its existence in A. This is the case since if an element z does not exist in B, then clearly it could not exist in A either.

On the other hand, existence of an element in A is not a necessary condition for its existence in B; it could certainly be the case that there exists an element y in B that does not exist in A.

Or, in other words: "A given polygon has four sides" is a **necessary** condition for "The given polygon is a square"; and "A given polygon is a square" is a **sufficient** condition for "The given polygon has four sides."

E.   The relationship between a necessary condition, a sufficient condition, and a conditional statement

Recall that in Chapter 8 we learned about the implication connective IF ..., THEN ... . Now, consider two statements p and q. Let us assume that p is a sufficient condition for q; in this case we can say that the conditional statement "IF p, THEN q" is TRUE. The equivalent conditional statement is: IF NOT q, THEN NOT p. In other words, q is a necessary condition for p.

We can also see the relationship between the sufficient condition and the conditional statement, and between the necessary condition and the conditional statement, in the Venn diagram shown above.

For example:

- "It is raining now" is a sufficient condition for "It is cloudy now" (p ["It is raining now"] is a sufficient condition for q ["It is cloudy now"]). Stated in the form of a conditional statement: "If it is raining, then it is cloudy" (IF p, THEN q).
- "It is cloudy now" is a necessary condition for "It is raining now" (q is a necessary condition for p). Stated in the form of a conditional statement: "If it is not cloudy, then it is not raining" (IF NOT q, THEN NOT p).

As stated above, there are many cases where a particular condition is necessary and sufficient. Such a case may be expressed as: p if and only if q ($p \leftrightarrow q$) (this was discussed in more detail in Sections 8.7g and 8.7h). For example, the condition: "Diagonals that bisect each other in a quadrilateral" is a necessary and sufficient condition for the quadrilateral to be a parallelogram. Thus, this condition could be used as a definition for the parallelogram (see more in Section G below).

F. The relationship between a necessary condition, a sufficient condition, and statements with the quantifiers ALL or ONLY

By looking at the Venn diagram shown above in Section D, we can see the connection that exists between necessary conditions and sufficient conditions on one hand, and the quantifiers ALL and ONLY on the other:

We see from the diagram that the set of squares is fully contained in the set of quadrilaterals. In other words: All squares are quadrilaterals. That is, it cannot be the case that there exists a square that is not a quadrilateral. This may also be expressed as follows: Only a polygon that is a quadrilateral could be a square. That is, it cannot be the case that there exists a polygon that is not four-sided that is a square.

Nevertheless, we have seen that "The polygon is a square" is a sufficient condition for "The polygon is a quadrilateral", and "The polygon is a quadrilateral" is a necessary condition for "The polygon is a square".

In general, the fact that p is a sufficient condition for q may be expressed as follows: ALL p ARE q.

The fact that q is a necessary condition for p may be expressed as: ONLY q IS p.

We use a Venn diagram to summarize the previous sections, drawing the following conclusions for all x:

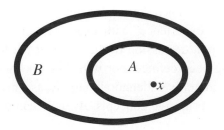

- $x \in A$ is a sufficient condition for $x \in B$; and $x \in B$ is a necessary condition for $x \in A$.
- IF $x \in A$, THEN $x \in B$ $(x \in A \rightarrow x \in B)$; and IF $x \notin B$, THEN $x \notin A$ $(x \notin B \rightarrow x \notin A)$.
- All elements that are in A are in B; and only an element that is in B is in A.

G. The meaning of "definition", and its relationship to necessary condition, sufficient condition, and to necessary and sufficient condition

G.1. Introduction — a mathematical theory

A mathematical theory is a collection of **truths** or proven assertions (that is, theorems, as they are conventionally called in mathematics). Every mathematical theory has a collection of **terms** used to express **concepts**, for which the relations between them or their special characteristics are the truths of that theory. (For example, in number theory, one of the terms is "prime number", and one of the truths is that there is an infinite number of prime numbers.) It is important that each term be **well-defined**, in order to guarantee that everyone who uses the theory has the same thing in mind when using the term. Thus, the definition must be precise and unambiguous. (For example, "prime number" is defined as "a natural number divisible by exactly two distinct natural numbers.")

As seen from this example, every definition is based on additional, underlying terms. (For example, the definition of a prime number is based on the terms "natural number" and "divisible.") Of course these prior terms must also be defined; their definitions are, in turn, based on definitions that preceded them. This regression of underlying terms must terminate at some point, in order to base a mathematical theory on a solid foundation, and in order to build it "bottom up." The fundamental concepts must be simple and straightforward, in order to assure that all users of the theory feel that they understand their meaning. These concepts are referred to as **primitive concepts**. For example, a child learns that a tomato is red and a cucumber is green. That child is not provided with a definition of "red", but learns the concept "red" from experience, with examples and counterexamples. Two children could engage in a discussion of the color of the tomato, using the term "red" without it having been defined, nor will they ask "what is red?" but rather they will accept it as commonly understood; namely, as a primitive term. The three conventionally accepted primitive concepts of Euclidean geometry are the point, the line, and the plane.

The meanings of primitive concepts are determined by the relations among them. These basic relations are called **axioms**, accepted as fundamentally true without proof. For example, in Euclidean geometry the following assertion is accepted as a fundamental truth: Between every two points there is one and only one straight line.

Game rules could illustrate the concept of axioms: the players of a game accept the rules without question, and play according to the rules without doubting them. A change in the rules of the game brings about a transformation to another game, which may be played according to the new rules, if accepted.

In summary, the cornerstones of any mathematical theory are its primitive concepts and its axioms. After laying the foundation, the theory may be built up upon it. This is accomplished by answering the question: "What else is true in a 'world' in which the primitive concepts and the given axioms are accepted?" From here, the theory is built up in two ways:

a. Definition of new terms to express new concepts based on the primitive concepts or on other concepts already defined based on them.
b. Proofs of new truths (called **theorems**), based on the axioms or theorems that have already been proven, using the terminology of the theory and the basic **rules of inference** (see Chapter 8).

The primitive concepts and the system of axioms, together with the newly defined concepts and the proven conclusions that follow from them, form **a mathematical theory**. A mathematical theory strives for the most solid foundation possible. Therefore, an axiomatic system must uphold three requirements:

1. Consistency — contradictory assertions cannot be deduced using the rules of inference from the axioms and from the theorems derived from them.
2. Independence — there does not exist an axiom that can be derived from other axioms (a minimal system).
3. Completeness — it is possible to prove or disprove every assertion that can be expressed within the theory based on its axiomatic system.

G.2. Definitions of terms

The definition of a term is a description of a concept represented by the term. As stated earlier, a definition must be precise and unambiguous in order for the idea associated with the defined term to be conceptualized uniformly in the minds of different people who use it. The definition must include only previously defined terms as well as basic conjunctions of language.

Why must terms be defined? — Definitions in mathematics allow for precise communication using terms as a kind of shortcut instead of having to describe explicitly mathematical concepts or objects in question again and again.

As shown in Chapter 2, the decision as to whether a certain geometric shape is a quadrilateral depends only on how a "quadrilateral" is defined.

What must a mathematical definition satisfy?

— According to Aristotle (384–322 B.C.E), a mathematical definition must uphold the following four conditions:

— **Criterion of Hierarchy** — Description of a new concept as a special case of a more general concept defined in a previous term, by indicating one or more properties that make this new concept a special case.

For example: A **right angle** is an **angle** in which the two adjacent sides are perpendicular to one another.

The general concept is "angle," and the property that uniquely identifies the new concept is that "two adjacent sides are perpendicular to one another." The new concept is "right angle."

— **Criterion of Existence** — A definition describes a concept, but does not establish the existence of an object that realizes that concept. Thus it is necessary to prove existence of an object that satisfies the definition; otherwise there is no need for the mathematical definition.

For example:

• We would like to define the concept "trisquacle" as a three-dimensional object whose projections are a triangle, a square (or rectangle) and a circle. Does such an object exist? The answer is "Yes". An example of this is a tube of toothpaste. So there is justification for introducing a new term to represent the new concept.

(From: https://www.pinterest.com/pin/492651646713636901/?lp=true)

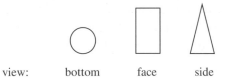

| view: | bottom | face | side |

- We would like to define an "exterior-median triangle" as a triangle whose medians intersect outside the triangle. Does such a triangle exist? — The answer is "No." In every triangle, the point of intersection of the medians is interior to the triangle. Thus, mathematically there is no need to define an "exterior-median triangle." Nonetheless, there is didactic importance to raising this possibility for discussion, in order to arrive at the conclusion that the definition of a new term to represent a non-existing concept is unnecessary.

— **Criterion of Equivalence** — Sometimes a particular concept can be defined in two (or more) ways, that is, with different properties. Let us denote these two ways as A and B. Two definitions are considered **equivalent** if and only if A can be selected as the definition of the concept and the properties in B can be proven from it, and vice versa — if B can be selected as the definition of the concept and the properties in A can be proven from it.

Examples:

- The concept parallelogram can be defined in the following ways:

  A. A quadrilateral with two pairs of parallel sides.
  B. A quadrilateral in which both pairs of opposite sides are of equal length.
  C. A quadrilateral in which one pair of opposite sides is parallel and of equal length.

All three of these definitions are equivalent, because it is possible to accept **each** of them as a definition and then to prove the remaining two.

— **Criterion of Minimality** — Euclid (365–325 B.C.E) presented an additional criterion. According to his requirement, the definition should not include properties beyond the minimum required to uphold the concept.

For example:

- The definition "A rectangle is a quadrilateral with four right angles" does not satisfy this requirement, since it suffices to define a rectangle as having three right angles because the sum of interior angles in every convex quadrilateral is 360°, and the size of the fourth angle is thus determined by the given size of the other three.

This criterion takes into account aesthetic-philosophical considerations more than logical ones, as a non-minimal definition does not create a contradiction in the mathematical theory. Still, for didactic reasons of clarity, it is often preferable to choose a non-minimal definition. For example, students may find it easier to accept a rectangle as a quadrilateral with four right angles rather than one with three.

G.3. Definition — a necessary and sufficient condition

In view of the discussion above, a definition is, in effect, a necessary and sufficient condition for a mathematical object to have a particular name.

Examples:

- The definition of an even number is: An integer divisible by 2. That is, a necessary and sufficient condition for an integer to be even is that the number is divisible by 2. Alternatively, an even number may be defined as an integer whose units digit is 0, 2, 4, 6 or 8. That is, a necessary and sufficient condition for an integer to be even is that its units digit is 0, 2, 4, 6 or 8. (The two alternative definitions are equivalent. Why?)

The equivalence of two assertions may be expressed using "if and only if" (iff).

For example:

- A number is even iff its units digits is one of: 0, 2, 4, 6, 8.
- A quadrilateral has a pair of opposite sides that are both parallel and equal iff its diagonals bisect each other.

Alternatively, as stated above, these assertions may be expressed using necessary and sufficient conditions.

9.5. Review of Answers to Worksheet 9b

After students have become familiar with the logical background, it is advised to ask them to go back and review their answers to Worksheet 9b and correct them if necessary.

9.6. Worksheets 9c and 9d: Necessary Conditions, Sufficient Conditions and Necessary and Sufficient Conditions

Worksheets 9c and 9d address the relationship between conditional statements and quantifiers on one hand, and necessary conditions and sufficient conditions on the other.

Remarks:

- This worksheet may be completed as individual practice, in pairs or in small groups, with group discussion encouraged.
- Upon completion of the worksheet, it is recommended that the students present their answers to the class to initiate a whole class discussion of the solutions.

Worksheet 9c and Proposed Solutions

In Worksheet 7d we used quantifiers to analyze the conversation between Alice and the Cheshire cat about people who live in Wonderland. Below we recall part of the conversation between them (page 76):

"In *that* direction," the Cat said, waving its right paw round, "lives a Hatter: and in *that* direction," waving the other paw, "lives a March Hare. Visit either you like: they're both mad."

"But I don't want to go among mad people," Alice remarked. "Oh, you can't help that," said the Cat: "we're all mad here. I'm mad. You're mad."

"How do you know I'm mad?" said Alice.

"You must be," said the Cat, "or you wouldn't have come here."

Which of the statements below is expressed in the excerpt above?

1. A necessary condition for someone to be mad is that they come to Wonderland.

2. A sufficient condition for someone to be mad is that they come to Wonderland.

3. In order for someone to come to Wonderland it is necessary for them to be mad.

4. In order for someone to come to Wonderland it is sufficient for them to be mad.

In order to find the right answer, we translate the excerpt above to a conditional statement.

Actually, the Cat says to Alice that if she were not mad, then she would not have come to Wonderland. Or, equivalently but expressed in the positive: If Alice came to Wonderland, then she is mad. In other words, "coming to Wonderland" is a sufficient condition for "being mad," and "being mad" is a necessary condition for "coming to Wonderland."

Therefore statements 2 and 3 are the statements that are expressed in the above excerpt. (Note that they are expressed in different ways, and both expressions should be attempted for both statements.)

Alternatively, the content of the excerpt could be expressed using the quantifier ALL as follows (as indicated in the solution to Worksheet 7d above as well):

> All those who come to Wonderland are mad.
> Alice came to Wonderland.
> Conclusion: Alice is mad.

Here, too, we see that statements 2 and 3 are the ones expressed in the above excerpt.

Worksheet 9d and Proposed Solutions

In Worksheet 3d we read a part of a conversation between Alice and the caterpillar. The caterpillar noticed that Alice was troubled by her height (which was, you may recall, only 3 inches tall), and it therefore suggested that she take a nibble of the mushroom on which it was sitting. The caterpillar promised her that a bite of the mushroom would make her taller. Alice nibbled on the mushroom, but evidently ate too much of it, since she grew so much that she started to feel that she could bend her neck easily in any direction, like a serpent. Because of her height, Alice's head reached the treetops, where she scared a pigeon that had flown into her face and was beating her violently with its wings. The Pigeon was certain that Alice was a serpent who had come to eat its eggs. Alice tried to convince the Pigeon that she was not a serpent. Below is an excerpt from the conversation between them (pages 61–62):

"But I'm *not* a serpent, I tell you!" said Alice. "I'm a—— I'm a ——"
"Well! *What* are you?" said the Pigeon. "I can see you're trying to invent something!"

"I — I'm a little girl," said Alice, rather doubtfully, as she remembered the number of changes she had gone through that day. "A likely story indeed!" said the Pigeon in a tone of the deepest contempt. "I've seen a good many little girls in my time, but never *one* with such a neck as that! No, no! You're a serpent; and there's no use denying it. I suppose you'll be telling me next that you never tasted an egg!"
"I *have* tasted eggs, certainly," said Alice, who was a very truthful child; "but little girls eat eggs quite as much as serpents do, you know."
"I don't believe it," said the Pigeon; "but if they do, why then they're a kind of serpent, that's all I can say."

---

1. Which of the following statements describes what the Pigeon said? (There may be more than one right answer).

    a. If a creature is a serpent, then it has a long neck.

    The Pigeon did not say that. The Pigeon identifies the neck but not the serpent. There could be serpents with necks that are different from the strange neck that brought the Pigeon to the conclusion that Alice is a serpent. But whoever has a neck like Alice's is identified by the Pigeon as a serpent.

(b.) If a creature has a long neck, then the creature is a serpent.

The Pigeon said that. When the Pigeon saw Alice's neck, it said: "You're a serpent; and there's no use denying it." In other words, if a creature has a neck like that, then the pigeon considers it a serpent.

c. Every serpent has a long neck.

The Pigeon did not say that. The Pigeon is not talking about serpents but about creatures that have necks like Alice's.

(d.) Only serpents have long necks.

The Pigeon said that. The Pigeon said that it had seen a good many little girls in its time, but never one with such a neck "as that." The Pigeon concludes that Alice must be a serpent, and it can be assumed that this follows from its overall experience that only serpents have long necks (but not necessarily all of them!).

e. A necessary condition for a creature to be a serpent is that it has a long neck.

The Pigeon did not say that. From what the Pigeon said it cannot be deduced that every serpent has a long neck like Alice's. Evidently the Pigeon deduces that Alice is a serpent based on its overall experience that only serpents have such a neck (but the Pigeon does not state explicitly that the conclusion follows from this general observation).

(f.) A sufficient condition for a creature to be a serpent is that it has a long neck.

The Pigeon said that. The Pigeon saw Alice with the long neck, and this sufficed for it to deduce that Alice is a serpent.

g. Are the statements that you indicated as describing what the Pigeon said equivalent to one another?

We marked statements b, d and f. We observed mathematically that the following three statements are equivalent to one another: "If A, then B," "Only B is A," and "A is a sufficient condition for B."

2. Below are four sequences of assertions. For each one of them answer the two following questions:

   A. Does the sequence reflect what is described in the quote above?
   B. Is the sequence valid as far as its logical inference is concerned?

   (i)   Only serpents like to eat eggs.
         Alice likes to eat eggs.
         Conclusion: Alice is a serpent.

In order to make it easier to grasp, we express the first assertion as a conditional statement:

      If a creature likes to eat eggs, then it is a serpent.
      Alice likes to eat eggs.
      Conclusion: Alice is a serpent.

This sequence is valid, in accordance with the "modus ponens" rule of inference (Section 8.7F), and this sequence reflects what the Pigeon said.

   (ii)   All serpents like to eat eggs.
         Alice likes to eat eggs.
         Conclusion: Alice is a serpent.

Here, too, it is easier to answer the question by expressing the first assertion as a conditional statement:

      If the creature is a serpent, then it likes to eat eggs.
      Alice likes to eat eggs.
      Conclusion: Alice is a serpent.

This sequence is invalid (as indicated in the discussion on the logical fallacy of "accepting the consequent" in Section 8.7G), and does not reflect what the Pigeon said.

   (iii)   A necessary condition for a creature to like to eat eggs is that it is a serpent.
         ALL little girls like to eat eggs.
         Conclusion: ALL little girls are serpents.

In this case we express the first two assertions as conditional statements.

> If a creature likes to eat eggs, then it is a serpent.
> If a creature is a little girl, then it likes to eat eggs.

It is easy to see that these two assertions, especially if we change their order, have the structure: If a, then b; If b, then c. Therefore, by transitivity (Section 6.4, C.3), the following may be inferred from them: If a, then c; or, in other words: If a creature is a little girl, then it is a serpent. The conclusion that appears in the question — All little girls are serpents — is equivalent to this. Therefore this sequence is valid and reflects what the Pigeon said.

(iv) A sufficient condition for a creature to like to eat eggs is that it is a serpent.
ALL little girls like to eat eggs.
Conclusion: ALL little girls are a serpents.

In this case, too, we express the first two assertions as conditional statements:

> If a creature is a serpent, then it likes to eat eggs.
> If a creature is a little girl, then it likes to eat eggs.

It is easy to see that these assertions have the structure: If a, then c; if b, then c. Therefore, it cannot be inferred that there is a relation between a and b. This sequence is invalid, and does not reflect what the Pigeon said.

3. Both of the following statements are TRUE in Wonderland:

(i) If a creature has a long neck, then it is a serpent and likes to eat eggs.
(ii) If a creature does not have a long neck, then it is not Alice.

Which of the following conclusions follows from the given statements? (There may be more than one right answer.)

Before we answer, let us designate the statements as follows:

   z: the creature has a long neck.
   n: the creature is a serpent.
   b: the creature likes to eat eggs.
   a: the creature is Alice.

We express the given statements using the indicated notation.

(i) $z \rightarrow (n \wedge b)$
(ii) $\sim z \rightarrow \sim a$

Statement (ii) is equivalent to: $a \rightarrow z$; therefore, by transitivity we get $[a \rightarrow z] \wedge [z \rightarrow (n \wedge b)] \equiv a \rightarrow (n \wedge b)$.

The given statements can also be expressed as a Venn diagram: Let N represent the set of serpents, let B represent the set of egg-eaters, let Z represent the long-necked creatures, and let A represent the set that has a single element — Alice.

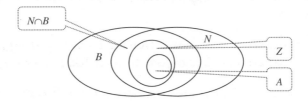

a.   If a creature does not like to eat eggs, then the creature does not have a long neck.

We express the assertion as follows: If $x \in \bar{B}$, then $x \in \bar{Z}$. Or, in other words: If $x \in Z$ then $x \in B$.

That is, $x \in Z$ is a sufficient condition for $x \in B$; we see this reflected in the diagram ($Z \subset B$). Therefore the assertion follows from the given information.

b.   If it is Alice, then it is a serpent.

We express the assertion as follows: If $x \in A$, then $x \in N$. Or, in other words, $x \in A$ is sufficient for $x \in N$, and we see this reflected in the diagram ($A \subset N$). Therefore, the assertion follows from the given information.

c.   If a creature does not have a long neck, then it is not a serpent.

We express the assertion as follows: If $x \in \overline{Z}$, then $x \in \overline{N}$. Or, in other words: If $x \in N$, then $x \in Z$.

That is, $x \in N$ is a sufficient condition for $x \in Z$. But in the Venn diagram we see that $Z \subset N$. Therefore, $x \in N$ is not a sufficient condition for $x \in Z$. (Moreover, it is a necessary condition.) Therefore, the assertion does not follow from the given information.

d.   All those who have a long neck are serpents.

We express the assertion as follows: If $x \in Z$, then $x \in N$. That is, $x \in Z$ is a sufficient condition for $x \in N$. We see this reflected in the diagram ($Z \subset N$); therefore, the assertion follows from the given information.

e.   In order to have a long neck, it is necessary to be a serpent.

We express the assertion as follows: $x \in N$ is a necessary condition for $x \in Z$. We see this reflected in the diagram ($Z \subset N$); therefore the assertion follows from the given information.

f.   If order to have a long neck it is necessary to be Alice.

We express the assertion as follows: $x \in A$ is a necessary condition for $x \in Z$. But from the diagram we see that $A \subset Z$. That is, $x \in A$ is not a necessary condition for this. Moreover, it is a sufficient condition for $x \in Z$. Therefore the assertion does not follow from the given information.

g.   In order to have a long neck it is necessary to like to eat eggs.

We express the assertion as follows: $x \in B$, is a necessary condition for $x \in Z$. We see this reflected in the diagram ($Z \subset B$); therefore the assertion follows from the given information.

h.   If a creature is a serpent and it is not Alice, then the creature does not have a long neck.

Let us examine the elements n, m, k. These three elements describe the different possibilities satisfying the requirements presented in the assumption of the conditional statement. It is TRUE that m, k ∉ Z, but n ∈ Z. Therefore the assertion does not follow from the given information.

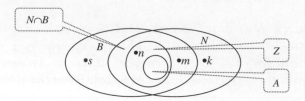

i.   In order to be a serpent, it suffices to have a long neck.

We express the assertion as follows: x ∈ Z is a sufficient condition for x ∈ N. We see this reflected in the diagram (Z ⊂ N); therefore the assertion follows from the given information.

j.   If a creature does not have a long neck, and it likes to eat eggs, then it is not Alice.

Let us take a look at the element m, indicated in the diagram of case h above. This element satisfies the requirements in the assumption of the conditional statement.

Another way to present this is by stating that x ∈ Z ∩ N is equivalent to x ∈ Z (since Z ⊂ N). Since x ∈ Z is a necessary condition for x ∈ A, we conclude that the assertion follows from the given information.

k.   There do not exist serpents that like to eat eggs and that are not long-necked.

Let us take a look at the element m, indicated in the diagram of case h above. This element represents a creature that is a serpent and likes to eat eggs, but is not long-necked. Therefore the assertion does not follow from the given information.

l.   In order not to be long-necked, it suffices not to be a serpent.

We express the assertion as follows: x ∈ N̄ is a sufficient condition for x ∈ Z̄ . Or, in other words, x ∈ Z is a sufficient condition for x ∈ N. We see this reflected in the diagram (Z ⊂ N); therefore the assertion follows from the given information.

m. All creatures that have a long neck must be Alice.

We express the assertion as follows: $x \in Z$ is a sufficient condition for $x \in A$. But from the diagram we see that $A \subset Z$. That is, $x \in Z$ is not a sufficient condition for $x \in A$; moreover, it is a necessary condition for it. Therefore the assertion does not follow from the given information.

n. Only a creature that is long-necked and likes to eat eggs is a serpent.

We express the assertion as follows: $x \in Z \cap B$ is a necessary condition for $x \in N$. But $x \in Z \cap B$ is equivalent to $x \in Z$ (since $Z \subset B$). Therefore, the assertion can be expressed as follows: $x \in Z$ is a necessary condition for $x \in N$. But in the diagram we see that $x \in Z$ is not necessary for $x \in N$. (Moreover, it is a sufficient condition for it.) Therefore the assertion does not follow from the given information.

9.7. Worksheets 9e and 9f: Summary Exercise for Necessary and Sufficient Conditions

The summary exercise is composed of two Worksheets: 9e and 9f.

Remarks:

- Worksheet 9e is to be completed as individual practice by students. In Worksheet 9f (which appears in the students' workbook only) students will compare their current answers with the answers they gave to the same questions at the start of the chapter (Worksheet 9a).
- Upon completion of the summary exercise, it is recommended that a whole class discussion be held in order to consider the changes that have taken place in students' understanding and perceptions through the course of this chapter, as well as to identify the particular difficulties encountered with the subject matter.

Worksheet 9e and Proposed Solutions

Answer the questions in this worksheet in their entirety, providing as much detail as possible. If necessary, you may indicate: "I did not understand the question, therefore I have not answered it." Make sure not to go back to Worksheet 9a before you complete your work on this worksheet.

In math class the students in Wonderland learned about a geometric shape called a *quantagon*. In their textbook the following statement appeared:

"In a *quantagon*, every two opposite sides are *quantivalent* to one another."

Which of the following assertions are TRUE and which are FALSE (provide an explanation for each case):

Before answering the question, we draw a Venn diagram describing the information about the quantagon.

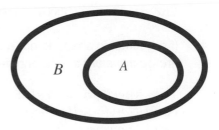

A — quantagons.
B — Geometric shapes in which any two opposite sides are quantivalent.

We express the properties of the quantagon in several ways, all of which are equivalent to one another.

a. *If a geometric shape is a quantagon, then all its opposite sides are quantivalent to one another.*
b. *If not all pairs of opposite sides of a geometric shape are quantivalent to one another, then the shape is not a quantagon.*
c. *In every quantagon every two opposite sides are quantivalent to one another.*
d. *Only geometric shapes all the opposite sides of which are quantivalent to one another are quantagons.*
e. *A sufficient condition for all pairs of opposite sides of a geometric shape to be quantivalent to one another is that the shape is a quantagon.*
f. *A necessary condition for a geometric shape to be a quantagon is that every two of its opposite sides are quantivalent to one another.*

1. The condition that every two opposite sides are *quantivalent* is a sufficient condition for a geometric shape to be a *quantagon*.

    The assertion is FALSE. Opposite sides that are quantivalent to one another is a necessary condition for a geometric shape to be a quantagon (as stated in property f). There might be other shapes having this property.

2. In every *quantagon* every two opposite sides are *quantivalent* to one another.

    This assertion is TRUE (as stated in property c).

3. If in a particular geometric shape not every two opposite sides are *quantivalent* to one another, then the shape is not a *quantagon*.

    The assertion is TRUE (as stated in property b).

4. If a geometric shape is not a *quantagon*, then not every two opposite sides are *quantivalent* to one another.

    The assertion is FALSE. "Being a quantagon" is a sufficient condition for all pairs of opposite sides to consist of two quantivalent sides; however it is not a necessary condition. It is certainly possible that there exist other geometric shapes that are not quantagons for which all pairs of opposite sides consist of two quantivalent sides.

5. In order for a geometric shape to be a *quantagon*, it is necessary that every two opposite sides are *quantivalent* to one another.

    The assertion is TRUE (as stated in property e).

Chapter 10

Comprehensive Summary and Review

IF Alice had found ONLY her way into the garden, she would NOT have met ALL the creatures that EXIST in Wonderland, NOR the White Rabbit OR the Queen of Hearts

This chapter is a summary and review of all topics of the course "Introduction to Mathematical Logic" that appear in this book. It includes a table summarizing all truth tables of the logical connectives discussed; a table comparing expression of statements as conditional statements, as statements with quantifiers and as necessary and sufficient conditions; and a table summarizing the various types of inference addressed. The chapter also introduces ten worksheets with exercises intended to allow the student to review all the topics presented in the course as an integrated unit. The excerpts that appear at the opening of each of the worksheets in this chapter come from Chapters 2 through 12 of *Alice's Adventures in Wonderland*.

Summary Table — Truth Tables of the Logical Connectives

p and q represent simple statements,
$\lor$ symbolizes OR,
$\land$ symbolizes AND,
$\rightarrow$ symbolizes IF ..., THEN ...

p	q	$p \lor q$	$p \land q$	$p \rightarrow q$
T	T	T	T	T
T	F	T	F	F
F	T	T	F	T
F	F	F	F	T

**Expression Comparison Table — Statements with Quantifiers,
Conditional Statements, and Necessary and Sufficient Conditions**

Statements with Quantifier Representation P and Q represent sets	Conditional Statement Representation p and q represent simple statements	Representation as Necessary and Sufficient Conditions p and q represent simple statements
FOR ALL $x \in P$, $x \in Q$	IF p, THEN q $(p \rightarrow q)$	p is a sufficient condition for q
ONLY $x \in Q$ is $x \in P$	IF $\sim q$, THEN $\sim p$ $(\sim q \rightarrow \sim p)$	q is a necessary condition for p
FOR ALL $x \in P$, $x \in Q$, and ONLY $x \in P$ is $x \in Q$	p IF AND ONLY IF q $(p \leftrightarrow q)$	p is a necessary and sufficient condition for q
Description using Venn diagrams 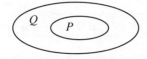		

Summary Table — Types of Inference

Types of Inference	Structure of Inference	Remarks
Modus Ponens	p and q represent simple statements (i) IF p, THEN q (ii) p Conclusion: q	The basic form of inference
Modus Tollens	p and q represent simple statements (i) IF p, THEN q (ii) $\sim q$ Conclusion: $\sim p$	The alternate (equivalent) basic form of inference
Affirming the consequent	p and q represent simple statements (i) IF p, THEN q (ii) q There is insufficient information to infer p and there is insufficient information to infer $\sim p$.	Inferring p is a common error
Denying the antecedent	p and q represent simple statements (i) IF p, THEN q (ii) $\sim p$ There is insufficient information to infer q and there is insufficient information to infer $\sim q$.	Inferring $\sim q$ is a common error

Worksheet 10a and Proposed Solutions

Worksheet 1c recounted how the White Rabbit dropped its fan and white kid gloves. Alice decided to help the White Rabbit hunt for the lost items, and arrived at the Rabbit's house. She entered the tidy little room, and found a fan and two or three pairs of white kid gloves. Her eye also fell upon a little bottle. There was no label this time, but nevertheless Alice was curious what would happen if she would drink from it. She did hope it would make her grow large again, for really she was quite tired of being such a tiny little creature. Before she had drunk half the bottle, she found that she had grown so that there was no longer room for her in the Rabbit's house. Alice needed to kneel down on the floor, and she put one arm out of the window, and one foot up the chimney.

The White Rabbit gathered many creatures around the house, and together they tried to come up with a solution for getting Alice out of the house. All their attempts proved unsuccessful. The following is a description that appears in *Alice's Adventures in Wonderland* (pages 44–45):

> "We must burn the house down!" said the Rabbit's voice. And Alice called out as loud as she could, "If you do, I'll set Dinah at you!"
> There was a dead silence instantly, and Alice thought to herself "I wonder what they will do next! If they had any sense, they'd take the roof off." After a min-
> ute or two they began moving about again, and Alice heard the Rabbit say "A barrowful will do, to begin with."
>
>  "A barrowful of *what*?" thought Alice. But she had not long to doubt, for the next moment a shower of little pebbles came rattling in at the window, and some of them hit her in the face. "I'll put a stop to this," she said to herself, and shouted out "You'd better not do that again!" which produced another dead silence.
>
> Alice noticed with some surprise that the pebbles were all turning into little cakes as they lay on the floor, and a bright idea came into her head. "If I eat one of these cakes," she thought, "it's sure to make some change in my size; and, as it can't possibly make me larger, it must make me smaller, I suppose."

1. Write the conditional statements that appear in the text above.

   If you do, then I'll set Dinah at you.
   If they had any sense, then they'd take the roof off.
   If I eat one of these cakes, then it's sure to make some change in my size.

2. Each item in the table below contains two statements followed by a conclusion. Rewrite the given statements and the proposed conclusion using statement notation. Then examine the proposed conclusion considering the given statements, and under the question "**Does the proposed conclusion follow from the statements?**" indicate 'Yes,' 'No,' or 'it cannot be determined' (due to insufficient information). Explain your answer.

Item	Statements	Proposed conclusions	Statement notation	Does the conclusion follow from the statements?
A	(i) If the White Rabbit burns the house down, then Alice will set Dinah at the Rabbit. (ii) Alice sets Dinah at the Rabbit.	The Rabbit burns the house down.	$p \rightarrow q$ q $\therefore p$	*It cannot be determined.*
Your reasoning: *It is also possible that Alice sets Dinah at the Rabbit for a different reason.*				

Let us designate the statements as follows:

　　p: The Rabbit will burn the house down.
　　q: Alice sets Dinah at the Rabbit.

We construct a truth table for the logical connective IF ... THEN ...

p	q	$p \to q$
T	T	T
T	F	F
F	T	T
F	F	T

It is given that $p \to q$ is TRUE. The first, third and fourth rows of the truth table for the conditional connective correspond to this bit of information.

It is also given that q is a TRUE statement. This narrows the possibilities to the first and third rows.

From these two rows we see that statement p can be either TRUE or FALSE; therefore it cannot be determined whether or not the proposed conclusion presented in the table is TRUE.

It is also possible that Alice sets Dinah at the Rabbit for a different reason.

p	q	$p \to q$
T	T	T
T	F	F
F	T	T
F	F	T

Item	Statements	Proposed conclusions	Statement notation	Does the conclusion follow from the statements?
B	(i) If the Rabbit's friends had any sense, then they'd take the roof off. (ii) The Rabbit's friends don't take the roof off.	The Rabbit's friends have no sense.	$p \to q$ $\sim q$ ——— $\therefore \sim p$	Yes

Your reasoning:

$p \to q \equiv \sim q \to \sim p$

Let us designate the statements as follows:

　　p: The Rabbit's friends have some sense.
　　q: The Rabbit's friends take the roof off.

We construct a truth table for the logical connective IF ... THEN ...

p	q	$p \to q$
T	T	T
T	F	F
F	T	T
F	F	T

It is given that $p \rightarrow q$ is TRUE. This bit of information corresponds to the first, third and fourth rows of the truth table for the conditional connective.

It is also given that q is FALSE. This narrows the possibilities to the fourth row only.

From this row, we can see that statement p is FALSE. The proposed conclusion is therefore TRUE.

p	q	$p \rightarrow q$
T	T	T
T	F	F
F	T	T
F	F	T

We may also arrive at this conclusion based on the equivalence of the statements $\sim q \rightarrow \sim p$ and $p \rightarrow q$.

Item	Statements	Proposed conclusions	Statement notation	Does the conclusion follow from the statements?
C	(i) If the Rabbit's friends fill a barrow with pebbles, then they can throw pebbles at Alice. (ii) The Rabbit's friends do not fill a barrow with pebbles.	The Rabbit's friends cannot throw pebbles at Alice.	$p \rightarrow q$ $\sim p$ ───── $\therefore \quad \sim q$	It cannot be determined.

Your reasoning:
The information we have relates to p. We do not know what happens in the case of $\sim p$.

Let us designate the statements as follows:

    p: The Rabbit's friends fill the barrow with pebbles.
    q: The Rabbit's friends can throw pebbles at Alice.

We construct a truth table corresponding to the logical connective IF ..., THEN ....

p	q	$p \rightarrow q$
T	T	T
T	F	F
F	T	T
F	F	T

It is given that $p \rightarrow q$ is TRUE. This bit of information corresponds to the first, third, and fourth rows of the truth table for the conditional connective.

It is also given that p is FALSE. This narrows the possibilities to the third and fourth rows.

From these rows we see that statement q can be either TRUE or FALSE; therefore it cannot be determined whether or not the proposed conclusion presented in the table is TRUE.

p	q	$p \rightarrow q$
T	T	T
T	F	F
F	T	T
F	F	T

In principle no conclusions can be drawn regarding what happens when p is FALSE, since all the information provided was only for the case that p is TRUE.

Item	Statements	Proposed conclusions	Statement notation	Does the conclusion follow from the statements?
D	(i) If the pebbles turn into little cakes, then they may be eaten. (ii) Pebbles turn into cakes.	Pebbles may be eaten.	$p \rightarrow q$ p ___ $\therefore \quad q$	Yes

Your reasoning:
Given p, then necessarily q.

Let us designate the statements as follows:

p: Pebbles turn into little cakes.
q: Pebbles may be eaten.

We construct a truth table corresponding to the logical connective IF ..., THEN ....

p	q	$p \rightarrow q$
T	T	T
T	F	F
F	T	T
F	F	T

It is given that $p \rightarrow q$ is TRUE. This bit of information corresponds to the first, third and fourth rows of the truth table for the material conditional.

	p	q	$p \to q$
It is also given that p is a TRUE statement. This narrows the possibilities to the first row.	T	T	T
From this row we obtain that statement p must be TRUE.	T	F	F
	F	T	T
In principle, if $p \to q$ is a TRUE statement, and p is a TRUE statement, then q must be a TRUE statement.	F	F	T

3. Each item below contains two statements followed by a question.

    — Answer the question considering the two given statements, and determine which of the following answers the correct one is: "yes," "no," "There is insufficient information given to determine." Explain your answer.
    — Then, express the statement that appears in question form as a conclusion, and determine whether the two statements and the assertion are equivalent to Item B of Question 2.

    A. Both of the following statements are TRUE in Wonderland:
        (i)   In order for the Rabbit's friends to take the roof off, it is sufficient for them to have some sense.
        (ii) The Rabbit's friends have no sense.

        Question: Do the Rabbit's friends take the roof off?

        Yes/No/There is insufficient information given to determine.

Let us designate the statements as follows:

p: The Rabbit's friends have some sense.
q: The Rabbit's friends take the roof off.

Statement (i) is equivalent to the statement $p \to q$. Statement (ii) is equivalent to the statement ~p.

It cannot be determined whether the Rabbit's friends take the roof off (whether q holds). The information we have relates to p. We do not know what happens in the case of ~p.

B. Both of the following statements are TRUE in Wonderland:

(i) In order for the Rabbit's friends to take the roof off, it is sufficient for them to have some sense.
(ii) The Rabbit's friends take the roof off.

Question: Do the Rabbit's friends have any sense?

Yes/No/There is insufficient information given to determine.

Let us designate the statements as follows:

   p: The Rabbit's friends have some sense.
   q: The Rabbit's friends take the roof off.

Statement (i) is equivalent to the statement $p \rightarrow q$. Statement (ii) is equivalent to the statement q.

It cannot be determined whether the Rabbit's friends have any sense (whether p holds). It is possible that they take the roof off for another reason; not necessarily because they have any sense.

C. Both of the following statements are TRUE in Wonderland:
(i) In order for the Rabbit's friends to have any sense, it is necessary for them to take the roof off.
(ii) The Rabbit's friends don't take the roof off.

Question: Do the Rabbit's friends have any sense?

Yes/No/There is insufficient information given to determine.

Let us designate the statements as follows:

   p: The Rabbit's friends have some sense.
   q: The Rabbit's friends take the roof off.

Statement (i) is equivalent to the statement $p \rightarrow q$. The statement (ii) is equivalent to the statement $\sim q$.

The statement $p \rightarrow q$ is equivalent to the statement $\sim q \rightarrow \sim p$.
It is given that $\sim q$. Therefore $\sim p$ necessarily holds, namely, the Rabbit's friends have no sense.
This case is equivalent to that in Item B of Question 2.

D. Both of the following statements are TRUE in Wonderland:
(i) In order for the Rabbit's friends to take the roof off, it is necessary for them to have some sense.
(ii) The Rabbit's friends have no sense.

Question: Do the Rabbit's friends take the roof off?

Yes/No/There is insufficient information given to determine.

Let us designate the statements as follows:

> *p: The Rabbit's friends have some sense.*
> *q: The Rabbit's friends take the roof off.*

Statement (i) is equivalent to the statement $q \to p$. Statement (ii) is equivalent to the statement $\sim p$.

The statement $q \to p$ is equivalent to the statement $\sim p \to \sim q$.
It is given that $\sim p$. Therefore $\sim q$ necessarily holds, namely, the Rabbit's friends don't take the roof off.

4. Both of the following statements are TRUE in Wonderland:
(i) If pebbles turn into little cakes, then they may be eaten.
(ii) If Alice's size does not get smaller, then the pebbles may not be eaten.

For each of the following assertions, determine whether it follows from the two given statements. Explain your answer.

A. If the pebbles turn into cakes, then Alice's size gets smaller.
B. If the pebbles do not turn into cakes, then Alice's size does not get smaller.
C. If Alice's size does not get smaller, then the pebbles don't turn into cakes.

Let us designate the statements as follows:

> *p: Pebbles turn into cakes.*
> *q: Pebbles may be eaten.*
> *r: Alice's size gets smaller*

A. *The assertion is: $p \rightarrow r$. Is this assertion true? As per the given information, the statements $p \rightarrow q$ and $\sim r \rightarrow \sim q$ are TRUE. The second may also be written using its equivalent statement: $q \rightarrow r$. Thus, by transitivity, we obtain: $p \rightarrow q \rightarrow r$. That is, the statement $p \rightarrow r$ is also TRUE; therefore the assertion A follows from the two given statements.*

B. *The assertion is: $\sim p \rightarrow \sim r$. Does this assertion follow from the two given statements? It cannot be determined, since we only know what happens given p, but we cannot infer anything regarding the case of $\sim p$.*

C. *The assertion is: $\sim r \rightarrow \sim p$. Does this assertion follow from the two given statements? In Item A we saw that the statement $p \rightarrow r$ is TRUE; therefore its equivalent statement: $\sim r \rightarrow \sim p$, is also TRUE. Hence Assertion C follows from the two given statements.*

Worksheet 10b and Proposed Solutions

After Alice swallowed one of the cakes, she became small and ran out of the house. She tried to find her way into that lovely garden, but she was so small at that point, that she needed to find a way to grow to her right size again. She supposed that she ought to eat or drink something or other; but the great question was, what? Suddenly she saw that there was a large mushroom growing near her, about the same height as herself, and, when she had looked under it, and on both sides of it, and behind it, it occurred to her that she might as well look and see what was on the top of it. Here is the description from the book (page 48):

She stretched herself up on tiptoe, and peeped over the edge of the mushroom, and her eyes immediately met those of a large blue caterpillar, that was sitting on the top with its arms folded, quietly smoking a long hookah, and taking not the smallest notice of her or of anything else.

1. Both of the following statements are TRUE in Wonderland:

    (i) The caterpillar is not taking the smallest notice of Alice or of anything else.
    (ii) The caterpillar is sitting on a mushroom.

    Which of the following statements is TRUE (there may be more than one right answer):

    a. The caterpillar is sitting on a mushroom or is taking notice of Alice.

    b. The caterpillar is sitting on a mushroom and is taking notice of the Rabbit.

    c. The caterpillar is taking notice of a leaf or is taking notice of Alice.

Let us designate the statements as follows:

  p: The caterpillar is not taking notice of Alice.
  q: The caterpillar is not taking notice of anything.
  r: The caterpillar is sitting on a mushroom.

It is given that the compound statement p ∧ q is TRUE (rows 1 and 2), and that r is TRUE (rows 1, 3, 5 and 7).

The corresponding row in the table for both possibilities is row 1. That is, p, q, r are TRUE. This can also be inferred without using a truth table; since the compound statement p ∧ q is TRUE, its two components are necessarily TRUE as well.

	p	q	r	p ∧ q
1	T	T	T	T
2	T	T	F	T
3	T	F	T	F
4	T	F	F	F
5	F	T	T	F
6	F	T	F	F
7	F	F	T	F
8	F	F	F	F

A. The statement given is r ∨ ~p. Is this statement TRUE? As we have seen, r is TRUE. A compound statement containing the logical connective OR is TRUE if at least one of its two component simple statements is TRUE; therefore the given statement is TRUE.

B. The statement given is r ∧ ~q. Is this statement TRUE? r is a TRUE statement, and q is a TRUE statement. Therefore ~q is FALSE. A compound statement that contains the logical connective AND is a TRUE statement if both of its two component simple statements are TRUE; therefore the given statement is FALSE.

C. The statement given is: ~q ∨ ~p. Is this statement TRUE? p and q are TRUE statements. Therefore the statement ~p and the statement ~q are FALSE. A compound statement that contains the logical connective OR is a FALSE statement if both of its two component simple statements are FALSE; therefore the given statement is FALSE.

2. In Happyland the following statement is TRUE:

(i) The caterpillar is not taking notice of Alice or the caterpillar is not taking notice of anything else.

In Happyland the following statement is FALSE:

(ii) The caterpillar is sitting on a mushroom.

Which of the following statements is TRUE in Happyland (there may be more than one right answer):

a. The caterpillar is taking notice of Alice and the caterpillar is sitting on a mushroom.

b. The caterpillar is taking notice of a snail or the caterpillar is not sitting on a mushroom.

c. The caterpillar is not sitting on a mushroom and the caterpillar is taking notice of Alice.

Let us designate the statements as follows:

p: The caterpillar is not taking notice of Alice.
q: The caterpillar is not taking notice of anything.
r: The caterpillar is sitting on a mushroom.

It is given that p ∨ r is TRUE; therefore at least one of its two components — the simple statements p or q — is TRUE. Note that it is not known whether both statements are TRUE or just one of them is TRUE.

The statement r is FALSE.

A. *The statement given is: ~p ∧ r. Is this statement TRUE? A compound statement that contains the logical connective AND is TRUE if both of its two component simple statements are TRUE. The truth value of p is not known (and therefore the truth value of ~p is also not known). But because statement r is FALSE, the given compound statement is FALSE.*

B. *The statement given is: ~q ∨ ~r. Is this statement TRUE? A compound statement that contains the logical connective OR is TRUE if at least one of its two component simple statements is TRUE. The truth value of statement q is not known (and therefore the truth value of ~q is also not known). But statement r is FALSE, therefore ~r is TRUE. Therefore the given compound statement is TRUE.*

C. *The statement given is: ~r ∧ ~p. Is this statement TRUE? A compound statement that contains the logical connective AND is TRUE if both of its two component simple statements are TRUE. The truth value of p is not known (and therefore the truth value of ~p is also not known). Statement r is FALSE; therefore ~r is TRUE. But because it is not known whether ~p is TRUE or FALSE, the truth value of the given compound statement cannot be determined.*

Worksheet 10c and Proposed Solutions

After Alice finally attracted its attention, the caterpillar took the hookah out of its mouth and addressed her. Here is their conversation (pages 49–50):

"Who are *you*?" asked the Caterpillar.

This was not an encouraging opening for a conversation. Alice replied, rather shyly,

"I hardly know, sir, just at present — at least I know who I *was* when I got up this morning, but I think I must have been changed several times since then."

"What do you mean by that?" said the Caterpillar sternly. "Explain yourself!"

"I can't explain *myself*, I'm afraid, sir," said Alice, "because I'm not myself, you see."

'I don't see,' said the Caterpillar.

"I'm afraid I can't put it more clearly," Alice replied very politely, "for I can't understand it myself to begin with; and being so many different sizes in a day is very confusing."

"It isn't," said the Caterpillar.

1. The following three statements are TRUE in Wonderland:

    (i) If someone doesn't understand himself, then he cannot explain himself.

    (ii) If someone is many different sizes in a day, then he is confused.

    (iii) If someone is not confused, then he can explain himself.

    Which of the following statements follows from the given information (there may be more than one right answer)?

    A. If someone does not understand himself, then he is confused.

    B. If someone is many different sizes in a day or cannot explain himself, then he is confused.

C. If someone understands himself, then he can explain himself.

D. If someone is many different sizes in a day and cannot explain himself, then he is confused.

E. If someone is not confused, then he is not many different sizes in a day and can explain himself.

F. If someone can explain himself, then he is not confused.

G. If someone can explain himself, then he does not understand himself or he is not confused.

First let us designate the statements as follows:

    p: Someone does not understand himself.
    q: Someone cannot explain himself.
    r: Someone is many different sizes in a day.
    t: Someone is confused.

We analyze a portion of the conclusions using a truth table (a truth table is not needed for the remainder as it is simple enough to understand without it).

A. The statement given is: $p \to t$. Does this statement follow from the given information?

Based on the given information, the following statements hold:

$p \to q$, $r \to t$, $\sim t \to \sim q$. The statement $\sim t \to \sim q$ is equivalent to $q \to t$, and thus by transitivity, we obtain:

$p \to q \to t$. That is, $p \to t$. Therefore, the assertion follows from the given information.

B. *The statement given is: $(r \lor q) \to t$. Does this statement follow from the given information?*

Let us construct a truth table for the three given conditional statements, and mark only the rows for which all three of these are TRUE. We also add to our truth table the conditional statement whose truth value is in question.

p	q	r	t	$p \to q$	$r \to t$	$q \to t$	$r \lor q$	$(r \lor q) \to t$
T	T	T	T	T	T	T	T	T
T	T	T	F	T	F	F	T	F
T	T	F	T	T	T	T	T	T
T	T	F	F	T	T	F	T	F
T	F	T	T	F	T	T	T	T
T	F	T	F	F	F	T	T	F
T	F	F	T	F	T	T	F	T
T	F	F	F	F	T	T	F	T
F	T	T	T	T	T	T	T	T
F	T	T	F	T	F	F	T	F
F	T	F	T	T	T	T	T	T
F	T	F	F	T	T	F	T	F
F	F	T	T	T	T	T	T	T
F	F	T	F	T	F	T	T	F
F	F	F	T	T	T	T	F	T
F	F	F	F	T	T	T	F	T

It can be seen in the table that in the shaded rows, the truth value of $(r \lor q) \to t$ is TRUE. That is, the assertion follows from the given information.

C. The statement given is: $\sim p \to \sim q$. Does this statement follow from the given information?

It is given that: $p \to q$. We do not know what happens in the case of $\sim p$. Therefore we cannot say that the statement follows from the given information.

D. The given statement is: $(r \wedge q) \to t$. Does this statement follow from the given information?

Let us use the table that appears in the answer to Item B above, and select the rows for which the three given conditional statement are TRUE. We replace the two rightmost columns so that they correspond to the conditional statement whose truth value is in question.

p	q	r	t	$p \to q$	$r \to t$	$q \to t$	$r \wedge q$	$(r \wedge q) \to t$
T	T	T	T	T	T	T	T	T
T	T	F	T	T	T	T	F	T
F	T	T	T	T	T	T	T	T
F	F	T	T	T	T	T	F	T
F	F	F	F	T	T	T	F	T

We see in these rows that the truth value of the statement: $(r \wedge q) \to t$ is TRUE. That is, the assertion follows from the given information.

E. The given statement is: $\sim t \to (\sim r \wedge \sim q)$. Does this statement follow from the given information?

Let us use the table that appears in the answer to Item B above, and select the rows for which the three given conditional statement are TRUE. We replace the two rightmost columns so that they correspond to the conditional statement whose truth value is in question.

p	q	r	t	$\sim r$	$\sim t$	$\sim q$	$\sim r \wedge \sim q$	$\sim t \to (\sim r \wedge \sim q)$
T	T	T	T	F	F	F	F	T
T	T	F	T	T	F	F	F	T
F	T	T	T	F	F	F	F	T
F	F	T	T	F	F	T	F	T
F	F	F	F	T	T	T	T	T

We see in the table that in the selected rows, the truth value of the statement $\sim t \to (\sim r \wedge \sim q)$ is TRUE. That is, the assertion follows from the given information.

F. The statement given is: $\sim q \to \sim t$. Does this statement follow from the given information?

It is given that $q \to t$. We do not know what happens in the case of $\sim q$. Therefore we cannot say that the statement follows from the given information.

G. *The statement given is:* ~q → (p ∧ ~ t). *Does this statement follow from the given information?*

Let us use the table that appears in the answer to Item B above, and select the rows for which the three given conditional statement are TRUE. We replace the two rightmost columns so that they correspond to the conditional statement whose truth value is in question.

p	*q*	*r*	*t*	*~q*	*~t*	*p ∧ ~t*	*~q → (p ∧ ~t)*
T	*T*	*T*	*T*	*F*	*F*	*F*	*T*
T	*T*	*F*	*T*	*F*	*F*	*F*	*T*
F	*T*	*T*	*T*	*F*	*F*	*F*	*T*
F	*F*	*T*	*T*	*T*	*F*	*F*	*F*
F	*F*	*F*	*F*	*T*	*T*	*F*	*F*

We see in the table that in the selected rows, the truth value of the statement: ~q → (p ∧ ~ t) is sometimes TRUE and sometimes FALSE. That is, we do not know whether the assertion follows from the given information.

2. Each item in the table below contains one or two statements followed by a conclusion. Examine the proposed conclusion considering the two given statement(s), and under the question "Does the proposed conclusion follow from the statement(s)?" indicate 'Yes,' 'No,' or 'it cannot be determined' (due to insufficient information). Explain your reasoning.

Item	Statement(s)	Proposed conclusions	Does the conclusion follow from the statements?
1	A. All those who do not understand something, cannot express it. B. Alice does not understand what goes on in Wonderland.	Alice cannot express what goes on in Wonderland.	*Yes*

Your reasoning:

Let us denote the following:
A: People who don't understand things
B: People who cannot express things
a: Alice
A. If $x \in A$ then $x \in B$.
B. It is given that $a \in A$; therefore we conclude that $a \in B$.

Item	Statement(s)	Proposed conclusions	Does the conclusion follow from the statements?
2	A. There exist people who don't know who they are. B. There exist people who don't know how to express things.	There exist people who don't know how to express things and don't know who they are.	*It cannot be determined.*

Your reasoning:

Let us denote the following:

A: People who don't know who they are.

B: People who do not know how to express things.

Several states are possible

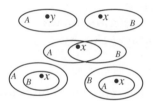

| 3 | There do not exist blue caterpillars that do not smoke hookahs. | All those who do not smoke a hookah are not caterpillars. | *Yes* |

Your reasoning:

The meaning of this statement is that all blue caterpillars smoke hookahs. In other words, if a caterpillar is blue, then it smokes a hookah. That is, whoever does not smoke a hookah is not a blue caterpillar.

Item	Statement(s)	Proposed conclusions	Does the conclusion follow from the statements?
4	A. Understanding things is a necessary condition for being able to explain them. B. All things that cannot be explained, cannot be expressed.	All things that are not understood cannot be expressed.	*Yes*

Your reasoning:

Let us denote the following:
A: Things that are understood.
B: Things that can be explained.
C: Things that can be expressed.
A. $x \in A$ is a necessary condition for $x \in B$. Or: If $x \in B$, then $x \in A$.
B. If $x \notin B$, then $x \notin C$. Or, equivalently: If $x \in C$, then $x \in B$.

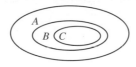

The proposed conclusion is: If $x \notin A$, then $x \notin C$.
Or, equivalently: If $x \in C$, then $x \in A$.

Item	Statement(s)	Proposed conclusions	Does the conclusion follow from the statements?
5	A. A necessary condition for explaining yourself is to know who you are. B. Only creatures who know how to explain themselves sit on mushrooms.	All creatures who know who they are sit on mushrooms.	*No*

Your reasoning:

Let us denote the following:
A: Creatures that can explain themselves.
B: Creatures that know who they are.
C: Creatures that sit on mushrooms.
A. $x \in B$ is a necessary condition for $x \in A$. Or: If $x \in A$, then $x \in B$.
B. Only $x \in A$ is $x \in C$. Or, equivalently: If $x \in C$, then $x \in A$.

Thus we obtain:

B. If $x \in C$, then $x \in A$.
A. If $x \in A$, then $x \in B$.

By transitivity, we obtain: If $x \in C$, then $x \in B$.
The corresponding Venn diagram is:

But the proposed conclusion is: If $x \in B$, then $x \in C$. We see therefore from the Venn diagram above that the conclusion is FALSE.

Worksheet 10d and Proposed Solutions

Alice continued talking to the caterpillar, who tried to understand what Alice wanted, as far as her height was concerned. Here is an excerpt from the book *Alice's Adventures in Wonderland* (pages 54–55):

> "What size do you want to be?" it asked.
>
> "Oh, I'm not particular as to size," Alice hastily replied; "only one doesn't like changing so often, you know."
>
> "I *don't* know," said the Caterpillar.
>
> Alice said nothing: she had never been so much contradicted in all her life before, and she felt that she was losing her temper.
>
> "Are you content now?" said the Caterpillar.
>
> "Well, I should like to be a *little* larger, sir, if you wouldn't mind," said Alice: "three inches is such a wretched height to be."
>
> "It is a very good height indeed!" said the Caterpillar angrily, rearing itself upright as it spoke (it was exactly three inches high).
>
> "But I'm not used to it!" pleaded poor Alice in a piteous tone. And she thought to herself, "I wish the creatures wouldn't be so easily offended!"

1. In Wonderland Forest live 600 creatures.

    300 of them are easily offended, 270 are 3 inches tall, and 210 are red caterpillars.

    120 are creatures that are easily offended and 3 inches tall.

    There are no red caterpillars that are easily offended.

    How many red caterpillars that are 3 inches tall live in Wonderland Forest?

Let us define our sets as follows:

A: The set of easily offended creatures
B: The set of creatures that are 3 inches tall
C: The set of red caterpillars

According to the given data:

$$n(A) = 300, \ n(B) = 270, \ n(C) = 210, \ n(A \cap B) = 120, \ n(C \cap A) = 0, \ n(C \cap B) = x$$

Since it is given that 120 creatures are in both sets A and B, and since the number of creatures in both sets B and C is unknown and designated x, the number of creatures in all three sets is: $300 + 270 + 210 - 120 - x = 600$. It follows from this that $x = 60$; that is, 60 creatures belong to the set $C - B$.

This can also be seen using a Venn diagram.

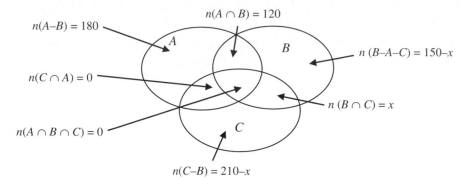

We obtain:

$$180 + 120 + 150\text{-}x + x + 210\text{-}x = 600 \Rightarrow x = 60$$

2. Both of the following statements are TRUE in Wonderland:

    (i)   All those who live in Wonderland are easily offended.

    (ii) If a creature doesn't live in Wonderland, then it is not a blue caterpillar.

Which of the following statements follows from the given information (there may be more than one right answer)?

A.   All those who are easily offended are blue caterpillars.

B.   There exist blue caterpillars that are easily offended.

C.   If someone is easily offended, then he lives in Wonderland.

D.   All those who live in Wonderland are blue caterpillars and are easily offended.

E.   There do not exist in Wonderland creatures that are not easily offended.

F.   Only someone who lives in Wonderland is easily offended.

G.   Only someone who is easily offended is a blue caterpillar.

H.   If someone lives in Wonderland, then he is easily offended and is a blue caterpillar.

I.   All those who are easily offended or are blue caterpillars, live in Wonderland.

Let us define our sets as follows:

    A: The set of creatures that live in Wonderland.
    B: The set of creatures that are easily offended.
    C: The set of blue caterpillars.

According to the given data:

(i) For all $x \in A$, $x \in B$. In other words: If $x \in A$, then $x \in B$.
(ii) If $x \notin A$, then $x \notin C$. In other words: If $x \in C$, then $x \in A$.

This can be seen using a Venn diagram:

A.   All those who are easily offended are blue caterpillars.

In other words: For all $x \in B$, $x \in C$.

The assertion does not necessarily follow from the given inform-ation. By transitivity, it follows from both bits of data that if $x \in C$, then $x \in B$. Or, in other words: For all $x \in C$, $x \in B$. This can also be seen by looking at the Venn diagram (in which $C \subset B$).

B.   There exist blue caterpillars that are easily offended.

In other words: There exists $x \in C$ such that $x \in B$.

This assertion necessarily follows from the given data (i), since from (ii) we gather that if it is a blue caterpillar, it is a creature that lives in Wonderland. This can also be seen by looking at the Venn diagram (in which $B \subset C$).

Note that it can even be inferred that every blue caterpillar is easily offended, since $C \subset B$.

C.   If someone is easily offended, then they live in Wonderland.

In other words: If $x \in B$, then $x \in A$.

The assertion does not necessarily follow from the given inform-ation. Assertion (i) maintains that if $x \in A$, then $x \in B$. But the reverse assertion (the converse) cannot be inferred from this.

D.   All those who live in Wonderland are blue caterpillar and are easily offended.

In other words: For all $x \in A$, $x \in B \cap C$.

The assertion does not necessarily follow from the given inform-ation. As we saw in Item A above, we can infer from the given information that $C \subset B$; therefore, $B \cap C = C$. Therefore, the given assertion is equivalent to the assertion: For all $x \in A$, $x \in C$, or, in other words: If $x \in A$, then $x \in C$. This assertion is the reverse (the converse) of the one in (ii), which asserts that: If $x \in C$, then $x \in A$. Since this is the case, we cannot infer that from assertion (ii).

E.   There do not exist in Wonderland creatures that are not easily offended.

In other words: All creatures in Wonderland are easily offended. Or: For all $x \in A$, $x \in B$. This is precisely the assertion given in (i).

F.   Only someone who lives in Wonderland is easily offended.

In other words: Only $x \in A$ is $x \in B$.

The assertion does not necessarily follow from the given information. This assertion is equivalent to the assertion: For all $x \in B$, $x \in A$. This assertion is the reverse of (i), which states that: For all $x \in A$, $x \in B$. We cannot infer from an assertion that its converse is TRUE.

G.   Only someone who is easily offended is a blue caterpillar.

In other words: Only $x \in B$ is $x \in C$.

This assertion is equivalent to the assertion: If $x \in C$, then $x \in B$, and it necessarily follows from the two given assertions, by transitivity. This can also be seen by looking at the Venn diagram (in which $C \subset B$).

H.   If someone lives in Wonderland, then he is easily offended and is a blue caterpillar.

In other words: If $x \in A$, then $x \in B \cap C$.

The assertion does not necessarily follow from the given information. Since $C \subset B$, $B \cap C = C$ holds TRUE as well; hence the assertion is: If $x \in A$, then $x \in C$, and it does not follow from the given information (explanation as per Item D).

I.   All those who are easily offended or are blue caterpillars, live in Wonderland.

In other words: For all $x \in B \cup C$, $x \in A$.

The assertion does not necessarily follow from the given information. Since $C \subset B$, $B \cup C = B$ holds TRUE as well. Therefore the given assertion is equivalent to the assertion: For all $x \in B$, $x \in A$, which does not follow from the given information (explanation as per Item C).

Worksheet 10e and Proposed Solutions

Alice arrived at the Duchess' house. The kitchen was all astir. The Duchess' cook put too much pepper in the soup, and everyone was sneezing, other than the cook and a large cat, which was sitting on the hearth and grinning from ear to ear. Alice asked the Duchess why the cat was grinning like that, and the Duchess answered that it was because it's a Cheshire cat. Here is their dialogue (page 69):

> "I didn't know that Cheshire cats always grinned; in fact, I didn't know that cats *could* grin."
>
>
>
> "They all can," said the Duchess; "and most of 'em do."
> "I don't know of any that do," Alice said very politely, feeling quite pleased to have got into a conversation.
> "You don't know much," said the Duchess; "and that's a fact."

1. Each item in the table below contains two statements followed by a conclusion. Rewrite the given statements and the proposed conclusion using statement notation. Then, examine the proposed conclusion considering the given statements, and under the question "**Does the proposed conclusion follow from the statements?**" Indicate 'Yes,' 'No,' or 'it cannot be determined' (due to insufficient information). Explain your answer in the corresponding column.

Item	Statements	Proposed conclusions	Statement notation	Does the conclusion follow from the statements?
A	Both of the following statements are TRUE in Wonderland: (i) If it is a Cheshire cat, then it is grinning. (ii) It is not a Cheshire cat.	The cat is grinning.	$p \to q$ $\sim p$ ————— $\therefore \quad q$	*It cannot be determined.*

Your reasoning:
The information we have relates to p.
We do not know what happens in the case of ~p.

Let us designate the statements as follows:

> *p: It is a Cheshire cat.*
> *q: The cat is grinning.*

We construct a truth table for the logical connective IF ... THEN ....

It is given that $p \to q$ is TRUE. This bit of information corresponds to the first, third and fourth rows of the truth table for the conditional connective.

p	q	$p \to q$
T	T	T
T	F	F
F	T	T
F	F	T

It is also given that ~p is a TRUE statement. That is, p is a FALSE statement. This narrows the possibilities to the third and fourth rows.

We mark these two rows. From these two rows we see that statement q can be either TRUE or FALSE; therefore it cannot be determined whether or not the proposed conclusion is TRUE. It is also possible that the cat is grinning for another reason.

Item	Statements	Proposed conclusions	Statement notation	Does the conclusion follow from the statements?
B	In Wonderland the following statement is TRUE: (i) If it is a Cheshire cat, then it is grinning. In Wonderland the following statement is FALSE: (ii) The cat is not grinning.	It is a Cheshire cat.	$p \rightarrow q$ q ————— $\therefore\ p$	*It cannot be determined.*

Your reasoning:
It is also possible that the cat is not Cheshire yet it is grinning.

Let us designate the statements as follows:

> *p: It is a Cheshire cat.*
> *q: The cat is grinning.*

We construct a truth table for the logical connective IF ..., THEN ... .

It is given that $p \rightarrow q$ is TRUE. This bit of information corresponds to the first, third and fourth rows of the truth table for the conditional connective.

It is also given that ~q is FALSE. That is, q is a TRUE statement. This narrows the possibilities to the first and third rows.

p	q	$p \rightarrow q$
T	T	T
T	F	F
F	T	T
F	F	T

We mark these two rows. From these two rows we see that statement p can be either TRUE or FALSE; therefore it cannot be determined whether or not the proposed conclusion is TRUE. It is also possible that it is not a Cheshire cat yet it is grinning.

2. In Wonderland the following statement is TRUE:

If a cat belongs to the Duchess, then it is a Cheshire cat.

Find the statements equivalent to it among the statements below:

A. If a cat is not a Cheshire cat, then it does not belong to the Duchess.

B. If a cat does not belong to the Duchess, then it is not a Cheshire cat.

C. A necessary condition for a cat to belong to the Duchess is that it is a Cheshire cat.

D. A sufficient condition for a cat not to be a Cheshire cat is that it belongs to the Duchess.

E. A necessary condition for a cat to be a Cheshire cat is that it belongs to the Duchess.

F. A necessary condition for a cat not to be a Cheshire cat is that it does not belong to the Duchess.

G. A sufficient condition for a cat not to be a Cheshire cat is that it does not belong to the Duchess.

Let us designate the statements as follows:

p: *The cat belongs to the Duchess.*
q: *It is a Cheshire cat.*
The given statement is: $p \rightarrow q$.

A. If a cat is not a Cheshire cat, then it does not belong to the Duchess.

The statement is: $\sim q \rightarrow \sim p$ *and it is equivalent to* $p \rightarrow q$. *From this we may infer that the statement given in this item is equivalent to the given true statement.*

B. If a cat does not belong to the Duchess, then it is not a Cheshire cat.

The statement is: $\sim p \rightarrow \sim q$. *Since this statement is not equivalent to* $p \rightarrow q$, *it follows that the statement given in this item is not equivalent to the given true statement.*

C.   A necessary condition for a cat to belong to the Duchess is that it is a Cheshire cat.

The statement in this item is: q is a necessary condition for p, or p is a sufficient condition for q. In expressing the statement as a conditional statement we obtain: $p \to q$, which is precisely the given true statement. Therefore the statement given in this item is equivalent to that in (i).

D.   A sufficient condition for a cat not to be a Cheshire cat is that it belongs to the Duchess.

The statement in this item is: A sufficient condition for ~q is p. In expressing the statement as a conditional statement we obtain: $p \to \sim q$; therefore the statement given in this item is not equivalent to the given true statement and even contradicts it.

E.   A necessary condition for a cat to be a Cheshire cat is that it belongs to the Duchess.

The statement in this item is: p is a necessary condition for q. Equivalently: q is a sufficient condition for p. Expressing the latter statement as a conditional statement, we obtain: $q \to p$, which is not equivalent to the reverse statement: $p \to q$. Therefore the statement given in this item is not equivalent to given true statement.

F.   A necessary condition for a cat not to be a Cheshire cat is that it does not belong to the Duchess.

The statement in this item is: ~p is a necessary condition for ~q. Or, equivalently: ~q is a sufficient condition for ~p. In expressing the latter as a conditional statement, we obtain: $\sim q \to \sim p$. This statement is equivalent to the statement: $p \to q$. Therefore, the statement given in this item is equivalent to the given true statement.

G. A sufficient condition for a cat not to be a Cheshire cat is that it does not belong to the Duchess.

The statement in this item is: ~p is a sufficient condition for ~q. In expressing the statement as a conditional statement we obtain: ~p → ~q. This statement is not equivalent to the statement p → q; therefore the statement in this item is not equivalent to the given true statement.

3. Both of the following statements are TRUE in Wonderland:

   (i) If it is a Cheshire cat, then the cat is sitting on the hearth and grinning.
   (ii) If it is not a Cheshire cat, then Alice doesn't know how to get into a conversation about it.

   Which of the following statements follows from the given information (there may be more than one right answer)?

   A. If a cat is not grinning, then it is not a Cheshire cat.

   B. If Alice knows how to get into a conversation about a cat, then the cat is sitting on the hearth.

   C. If it is not a Cheshire cat, then the cat is not sitting on the hearth.

   D. All Cheshire cats sit on the hearth.

   E. There do not exist cats which do not sit on the hearth that Alice doesn't know how to get into conversations about.

   F. A cat sitting on the hearth is a necessary condition for it to be a Cheshire cat.

   G. Alice's knowledge of getting into a conversation about cats is a necessary condition for it to be a Cheshire cat.

   H. If a cat is sitting on the hearth and Alice doesn't know much about the cat, then it is not a Cheshire cat.

   I. If a cat is not sitting on the hearth and the cat is grinning, then Alice doesn't know how to get into a conversation about the cat.

   J. There do not exist cats that sit on the hearth and grin, about which Alice doesn't know how to get into a conversation.

Let us define our sets as follows:

    A: The set of Cheshire cats.
    B: The set of cats that sit on the hearth.
    C: The set of grinning cats.
    D: The set of cats that Alice knows how to get into a conversation about.

Note: We defined D in this way so that we can relate it to information in (ii), and to be able to connect it to the information in (i).

According to the given data:

(i) *If $x \in A$, then $x \in B \cap C$.*
(ii) *If $x \notin A$, then $x \notin D$. In other words: If $x \in D$, then $x \notin A$.*

By transitivity, we obtain: If $x \in D$, then $x \in B \cap C$.
This can be seen using a Venn diagram:

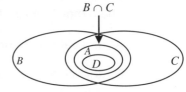

A. If a cat is not grinning, then it is not a Cheshire cat.

    In other words: If $x \notin C$, then $x \notin A$. Or, equivalently: If $x \in A$, then $x \in C$.

    The assertion necessarily follows from the given information. According to the given information: If $x \in A$, then $x \in B \cap C$. That is: If $x \in A$, then $x \in B$. And if $x \in A$, then $x \in C$. Thus we obtain that the statement in this item follows from that in statement (i). This can also be seen by looking at the Venn diagram (in which $A \subset C$).

B. If Alice knows how to get into a conversation about a cat, then the cat is sitting on the hearth.

    In other words: If $x \in D$, then $x \in B$.

    The assertion necessarily follows from the given information. As we have seen, if $x \in D$, then $x \in B \cap C$ by transitivity. It follows, therefore, similar to what was explained in Item A above, that if $x \in D$, then $x \in B$. Thus we obtain that the statement in this item follows from statements (i) and (ii). This can also be seen by looking at the Venn diagram (in which $D \subset B$).

C. If it is not a Cheshire cat, then the cat is not sitting on the hearth.

In other words: If $x \notin A$, then $x \notin B$. Or, equivalently: If $x \in B$, then $x \in A$.

The assertion does not necessarily follow from the given information. Assertion (i) *maintains that if $x \in A$, then $x \in B \cap C$. That is: If $x \in A$, then $x \in B$. But the converse assertion cannot be inferred from this.*

D. All Cheshire cats sit on the hearth.

In other words: For all $x \in A$, $x \in B$.

The assertion necessarily follows from the given information. Assertion (i) *maintains that for all $x \in A$, $x \in B \cap C$; therefore it necessarily follows that for all $x \in A$, $x \in B$. This can also be seen by looking at the Venn diagram (in which $A \subset B$).*

E. There do not exist cats which do not sit on the hearth that Alice doesn't know how to get into conversations about.

In other words: All cats that Alice doesn't know how to get into conversations about, sit on the hearth. Or: For all $x \notin D$, $x \in B$.

The assertion does not necessarily follow from the given information.

According to statement (i), *if Alice knows how to get into a conversation about a cat, then the cat is necessarily a Cheshire cat. But we do not have any information about cats that Alice does not know how to get into conversations about. They could be Cheshire cats, but they could also be another kind of cat. It is also known that all Cheshire cats sit on the hearth. But we do not know about other cats that they necessarily do not sit on the hearth. Perhaps they do, and perhaps they don't. Therefore, as indicated above, cats that Alice doesn't know how to get into conversations about (that do not belong to set D), could be Cheshire cats, and could be of another kind, and they could be on the hearth (belong to set B), but they could also not be on the hearth (not belong to set B).*

This can easily be seen by looking at the Venn diagram. For example,

$x \notin D$, $y \notin D$, but $x \notin B$ while $y \in B$. Therefore, the assertion does not necessarily follow from the given information.

F. A cat sitting on the hearth is a necessary condition for it to be a Cheshire cat.

In other words: $x \in B$ is a necessary condition for $x \in A$.

The assertion necessarily follows from the given information.

According to statement (i), $x \in B \cap C$ is a necessary condition for $x \in A$. From this in particular, it follows that belonging to B is a necessary condition for belonging to A. This can be seen by looking at the Venn diagram (in which $A \subset B$).

G. Alice's knowledge of getting into a conversation about cats is a necessary condition for it to be a Cheshire cat.

In other words: $x \in D$ is a necessary condition for $x \in A$.

The assertion does not necessarily follow from the given information.

Since according to statement (ii), $x \in D$ is a sufficient condition for $x \in A$. We have learned that if a condition is sufficient it does not imply that the condition is also necessary. It can be seen in the diagram that $D \subset A$.

H. If a cat is sitting on the hearth and Alice doesn't know how to get into a conversation about it, then it is not a Cheshire cat.

In other words, if $x \in B \cap \bar{D}$, then $x \notin A$.

The assertion does not necessarily follow from the given information.

We learn from statement (i) that every Cheshire cat sits on the hearth, but there could also be other kinds of cats on the hearth; that is, $x \in B$ could be a Cheshire cat but could also be another kind of cat.

We learn from statement (ii) that if Alice knows how to get into a conversation about a cat, then the cat is necessarily a

Cheshire cat. But we have no information about cats that Alice does not know how to get into a conversation about — they could be Cheshire cats, but they could also be another kind of cat. That is, $x \in \bar{D}$ could be a Cheshire cat, but it could be a different kind of cat.

Therefore, cats with both of these properties — sit on the hearth and Alice does not know how to get into a conversation about them — could be Cheshire cats, but it is also possible that they are not Cheshire cats.

A look at the Venn diagram can help in understanding of these relationships.

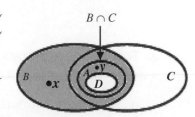

For example: $x \in B \cap \bar{D}$ $y \in B \cap \bar{D}$. But $x \notin A$, while $y \in A$.

Therefore, the assertion does not necessarily follow from the given information.

I. If a cat is not sitting on the hearth and the cat is grinning, then Alice doesn't know how to get into a conversation about the cat.

In other words: If $x \in \bar{B} \cap C$, then $x \in \bar{D}$.

The assertion necessarily follows from the given information.

According to statement (i), a cat that doesn't sit on the hearth is not a Cheshire cat. We infer from this that a cat that is not sitting on the hearth and is grinning, is not a Cheshire cat either.

According to statement (ii), a cat that is not a Cheshire cat is a cat that Alice doesn't know how to get into a conversation about.

By transitivity, it follows that a cat that doesn't sit on the hearth and is grinning is a cat that Alice doesn't know how to get into a conversation about. This is exactly what the given statement asserts.

This can also be seen in the diagrams shown below. In Diagram 1 the set shown in gray is $\bar{B}$. In Diagram 2 the set shown in gray *is $\bar{B} \cap C$. In Diagram 3 the set shown in gray is $\bar{D}$. It can be seen from Diagrams 2 and 3 that the set $\bar{B} \cap C$ is contained in the set $\bar{D}$, and this demonstrates the fact that the given statement follows from the two statements in (i) and (ii).*

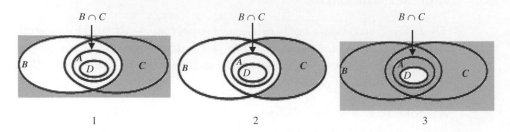

J. There do not exist cats that sit on the hearth and grin that Alice doesn't know how to get into a conversation about.

A statement equivalent to this is: For all cats that sit on the hearth and grin, Alice knows how to get into a conversation about them. Or, in statement notation: For all $x \in B \cap C$, $x \in D$.

The assertion does not necessarily follow from the given information. From statement (i) we infer (based on the explanation provided previously) that cats that sit on the hearth and grin could be one of two kinds — Cheshire cats, or cats that are not Cheshire cats. As far as cats that are not Cheshire cats, we know from (ii) that Alice doesn't know how to get into a conversation about them. But as far as Cheshire cats, we don't know for certain — there may be those about which Alice can get into a conversation, and there may be those about which she does not. Thus, by transitivity, the assertion given in this item that all those cats are necessarily ones that Alice can get into a conversation about doesn't necessarily follow from the given information.

This can be seen by looking at the diagram: $D \subset B \cap C$. For example: $x \in B \cap C$, $y \in B \cap C$; but $x \notin D$, while $y \in D$.

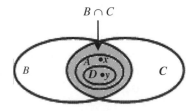

Worksheet 10f and Proposed Solutions

Recall that in Worksheet 7d and Worksheet 9c we analyzed the conversation between Alice and the Cheshire cat. The cat told her about the March Hare who goes mad in March. Alice arrived at the March Hare's house and found a table set out under a tree in front of the house, with the March Hare and the Hatter having tea at it. A Dormouse was sitting between them, fast asleep. When Alice sat down at one end of the table, the Hatter opened his eyes very wide and said "Why is a raven like a writing-desk?" Then, the following conversation took place between them (pages 83–84):

"Come, we shall have some fun now!" thought Alice.
"I'm glad they've begun asking riddles. — I believe I can guess that," she added aloud.
"Do you mean that you think you can find out the answer to it?" said the March Hare.

"Exactly so," said Alice.
"Then you should say what you mean," the March Hare went on.
"I do," Alice hastily replied;
"at least — at least I mean what I say — that's the same thing, you know."
"Not the same thing a bit!" said the Hatter. "Why, you might just as well say that 'I see what I eat' is the same thing as 'I eat what I see'!"
"You might just as well say," added the March Hare, "that 'I like what I get' is the same thing as 'I get what I like'!"
"You might just as well say," added the Dormouse, which seemed to be talking in his sleep, "that 'I breathe when I sleep' is the same thing as 'I sleep when I breathe'!"
"It *is* the same thing with you," said the Hatter; and here the conversation dropped.

1.   A. How can Alice's statement: "I mean what I say" be expressed as a conditional statement?

> *When Alice says: "I mean what I say," she's actually saying: "If I say something, then I mean it."*

     B. How can the Hatter's words be expressed as statements beginning with the words: "A necessary condition for..."?

> *The Hatter's words may be expressed as follows: "You might just as well say that 'a necessary condition for eating something is that I see it' is the same as 'a necessary condition for seeing something is that I am eating it'!"*

     C. How can the March Hare's words be expressed as statements that use the quantifier ALL?

> *The March Hare's words can be expressed as follows: "You might just as well say that 'ALL I get I like' is the same as 'ALL I like I get'!"*

     D. How can the Dormouse's words be expressed using statements containing negation?

> *The Dormouse's words can be expressed as follows: "You might just as well say that 'when I don't breathe I don't sleep' is the same as 'when I don't sleep I don't breathe'!"*

     E. How can the Hatter's words to the Dormouse be expressed as a conditional statement?

> *The Dormouse says to Alice: "You might just as well say that 'I breathe when I sleep' is the same as 'I sleep when I breathe'!" And the Hatter responds and says to the Dormouse that it is the same. Therefore the Hatter is saying, in effect, that the Dormouse could say: "I breathe if and only if I am sleeping."*

2. The following three statements are TRUE in Wonderland:

   (i) Having fun is a necessary condition for finding out the answers to riddles.

   (ii) Whoever cannot find out the answers to riddles doesn't participate in the Hatter's tea party.

   (iii) There exists a creature who is having fun who doesn't participate in the Hatter's tea party.

   Which of the following statements follows from the given information (there may be more than one right answer):

   A. Participation in the Hatter's tea party is a sufficient condition for the participant to be having fun.

   B. If someone doesn't find out the answers to riddles, then he is not having fun.

   C. All those who are having fun participate in the Hatter's tea party.

   D. Only someone having fun can find out the answers to riddles.

   E. Not having fun is a sufficient condition for not finding out the answers to riddles.

   F. There does not exist a creature that participates in the Hatter's tea party and is not having fun.

   G. There exists a creature that is having fun that cannot find out the answers to riddles.

   Let us define our sets as follows:

   A: The set of creatures that are having fun.
   B: The set of creatures that can find out the answers to riddles.
   C: The set of creatures that participate in the Hatter's tea party.

   According to the given data:

   (i) $x \in A$ *is a necessary condition for* $x \in B$. *In other words: If* $x \in B$, *then* $x \in A$.

   (ii) *For all* $x \notin B$, $x \notin C$. *In other words: If* $x \notin B$, *then* $x \notin C$. *Or: If* $x \in C$, *then* $x \in B$.

By transitivity we obtain: If $x \in C$, then $x \in A$.

(iii) There exists $x \in (A-C)$.

This can be seen using a Venn diagram:

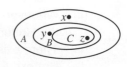

A. Participation in the Hatter's tea party is a sufficient condition for the participant to be having fun.

In other words: $x \in C$ is a sufficient condition for $x \in A$. Or, equivalently: If $x \in C$, then $x \in A$.

The assertion necessarily follows from the given information.

From statement (ii), we infer that if a creature participates in the tea-party $x \in C$, then the creature finds out the answers to riddles $x \in B$. From statement (i) we infer that if a creature finds out the answers to riddles $x \in B$, then it is having fun $x \in A$. Thus, by transitivity, the assertion in the given item follows from the given information.

This can also be seen in the Venn diagram (where $C \subset A$).

B. If someone doesn't find out the answers to riddles, then he is not having fun.

In other words: If $x \notin B$, then $x \notin A$. Or, equivalently: If $x \in A$, then $x \in B$.

The assertion does not necessarily follow from the given information. Assertion (i) maintains that if $x \in B$, then $x \in A$. But the converse assertion cannot be inferred from this.

This can also be seen in the diagram (where, for example, $x \in A$; $y \in A$, but $x \notin B$, while $y \in B$).

C. All those who are having fun participate in the Hatter's tea party.

In other words: For all $x \in A$, $x \in C$. Or, equivalently: If $x \in A$, then $x \in C$.

The assertion does not necessarily follow from the given information. Although we proved in Item A that if $x \in C$, then $x \in A$, the converse assertion cannot be inferred from this.

This can also be seen by looking at the Venn diagram (in which, for example, $x \in A$; $z \in A$, but $x \notin C$, while $z \in C$).

D. Only someone having fun can find out the answers to riddles.

In other words: Only $x \in A$ is $x \in B$. Or, equivalently: $x \in A$ is a necessary condition for $x \in B$. And this is exactly what is given in statement (i).

The assertion necessarily follows from the given information.

E. Not having fun is a sufficient condition for not finding out the answers to riddles.

In other words: $x \notin A$ is a sufficient condition for $x \notin B$. Or, equivalently: If $x \notin A$, then $x \notin B$, or: If $x \in B$, then $x \in A$. And this is exactly what is given in statement (i).

The assertion necessarily follows from the given information.

F. There does not exist a creature that participates in the Hatter's tea party and is not having fun.

In other words: There does not exist $x \in C$ such that $x \notin A$, or: For all $x \in C$, $x \in A$.
The assertion necessarily follows from the given information, as already seen in Item A.

G. There exists a creature that is having fun that cannot find out the answers to riddles.

In other words: There exists $x \in A$ such that $x \notin B$.

The assertion does not necessarily follow from the given information. According to (iii) *there exists a creature that is having fun that doesn't participate in the tea-party (There exists $x \in (A - C)$). Such a creature may find out the answers to riddles $x \in B$ and may not find out the answers to riddles $x \notin B$. There is no additional given information that could resolve this.*

This can also be seen by looking at the diagram: The elements x and y represent possibilities consistent with the given statement (iii). *The assertion given in this item corresponds to existence of such an element x, but the data does not allow us to negate the existence of an element such as y; thus the existence of x asserted in this item is not guaranteed.*

Worksheet 10g and Proposed Solutions

The Hatter was the first to break the silence by asking "What day of the month is it?" and taking his watch out of his pocket, and looking at it uneasily. Alice had been looking over the Hatter's shoulder with some curiosity. The watch seemed funny to Alice, because it told the day of the month, and not what o'clock it is. Alice remarked as such to the Hatter, and after a short conversation about the watch, the Hatter turned to Alice and asked her if she already guessed the solution to the riddle. Here is their conversation (page 87):

Alice sighed wearily. "I think you might do something better with the time," she said "than wasting it asking riddles with no answers."

"If you knew Time as well as I do," said the Hatter, "you wouldn't talk about wasting *it*. It's *him*."

"I don't know what you mean," said Alice.

"Of course you don't!" the Hatter said, tossing his head contemptuously.

"I daresay you never spoke to Time!"

"Perhaps not," Alice cautiously replied: "but I know I have to beat time when I learn music."

"Ah! that accounts for it," said the Hatter. "He won't stand beating. Now, if you only kept on good terms with him, he'd do almost anything you liked with the clock. For instance, suppose it were nine o'clock in the morning, just time to begin lessons: you'd only have to whisper a hint to Time, and round goes the clock in a twinkling! Half-past one, time for dinner!"

1. Let us define sets as follows:

   A: The set of hatters.

   B: The set of creatures that can beat time.

   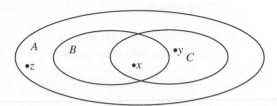

   C: The set of creatures that toss their heads contemptuously.

   We describe relations between the sets with a Venn diagram (see above). Now, let x, y and z be elements belonging to the sets as described in the diagram.

   How can the relations between the sets be expressed in four different statements using the connectives NOT and AND, and the quantifiers ALL (or FOR ALL), ONLY, and THERE EXISTS?

   One way to describe the information given in the diagram is:

   (i) *For all $x \notin A$, $x \notin B$. Or, in other words: If $x \in B$, then $x \in A$.*
   (ii) *Only $x \in A$ is $x \in C$. Or, in other words: If $x \in C$, then $x \in A$.*
   (iii) *There exists $x \in B \cap C$.*
   (iv) *There exists $y \in C - (B \cap C)$.*

   Thus, a possible answer is:

   (i) *All those who are not hatters cannot beat time.*
   (ii) *Only a hatter tosses his head contemptuously.*
   (iii) *There exist creatures that can beat time and toss their heads contemptuously.*
   (iv) *There exist creatures that toss their head contemptuously and cannot beat time.*

2.   A. Construct two assertions that follow necessarily from the statements you gave above.

     B. Construct two assertions that do not necessarily follow from the statements you gave above.

     C. Construct two assertions that contradict at least one of the statements you gave above.

     A. Example 1:

     Creatures that toss their head contemptuously are hatters.

     In other words: For all $x \in C$, $x \in A$. Or: If $x \in C$, then $x \in A$.

     The assertion necessarily follows from the given information. It can be seen in the diagram that $C \subset A$.

     Example 2:

     Only hatters are creatures that beat time and toss their heads contemptuously.

     In other words: Only $x \in A$ is $x \in B \cap C$. Or, If $x \in B \cap C$, then $x \in A$.

     The assertion necessarily follows from the given information. It can be seen in the diagram that $(B \cap C) \subset A$.

     Example 3:

     Being a hatter is a necessary condition for tossing one's head contemptuously.

     In other words: $x \in A$ is a necessary condition for $x \in C$. Or: If $x \in C$, then $x \in A$.

     The assertion necessarily follows from the given information. It can be seen in the diagram that $C \subset A$.

Example 4:

Not being a hatter is a necessary condition for not tossing one's head contemptuously.

In other words: $x \notin A$ is a necessary condition for $x \notin C$. Or: $x \in C$ is a sufficient condition for $x \in A$, or: If $x \in C$, then $x \in A$.

The assertion necessarily follows from the given information. It can be seen in the diagram that $C \subset A$.

Example 5:

There exists a hatter that can beat time.

In other words: There exists $x \in A$ such that $x \in B$.

The assertion necessarily follows from the given information. Element x that appears in the diagram and whose existence is guaranteed by statement (iii), is an example of this.

B. *Example 1:*

If someone is a hatter, then he can beat time.

In other words: If $x \in A$, then $x \in B$.

The assertion does not necessarily follow from the given information. It can be seen in the diagram that $B \subset A$; therefore if $x \in B$, then $x \in A$, but the converse assertion cannot be inferred from this.

Example 2:

If someone can beat time, then he tosses his head contemptuously.

In other words: If $x \in B$, then $x \in C$.

The assertion does not necessarily follow from the given information. From (iii) we know that there exists $x \in B \cap C$, but since $B \subset C$ is not given, the assertion cannot be inferred.

Example 3:

Only someone who can beat time is a hatter.

In other words: Only $x \in B$ is $x \in A$. Or: If $x \in A$, then $x \in B$.

The assertion does not necessarily follow from the given information (as explained in Item A).

Example 4:

Hatters toss their heads contemptuously.

In other words: For all $x \in A$, $x \in C$. Or: If $x \in A$, then $x \in C$.

The assertion does not necessarily follow from the given information. It can be seen in the diagram that $C \subset A$; therefore if $x \in C$, then $x \in A$, but the converse assertion cannot be inferred.

Example 5:

Tossing one's head contemptuously is a necessary condition for not being able to beat time.

In other words: $x \in C$ is a necessary condition for $x \notin B$. Or: If $x \notin B$, then $x \in C$.

The assertion does not necessarily follow from the given information. In the diagram we see that although the element y is an example of this, it is also possible that there exists an element like z.

C. Example 1:

Not being able to beat time is a sufficient condition for someone not being a hatter.

In other words: $x \notin B$ is a sufficient condition for $x \notin A$. If $x \notin B$, then $x \notin A$. The assertion contradicts the given information. The element y that appears in the diagram, whose existence is guaranteed from statement (iv), is an example of an element that contradicts the condition: If $x \notin B$, then $x \notin A$.

Example 2:

There does not exist a creature that is a hatter who tosses his head contemptuously and can't beat time.

In other words: There does not exist $x \in A \cap C$ such that $x \in \bar{B}$.

The assertion contradicts the given information. Since $C \subset A$, $A \cap C = C$ necessarily holds. Therefore the assertion may be expressed as follows: There does not exist $x \in C$ such that $x \in \bar{B}$. The element y that appears in the diagram, whose existence is guaranteed by statement (iv), *is an example of an element that contradicts the assertion.*

Example 3:

There exist creatures that can beat time and toss their heads contemptuously who are not hatters.

In other words: There exists $x \in B \cap C$ such that $x \notin A$.

The assertion contradicts the given information. It can be seen in the diagram that $(B \cap C) \subset A$; therefore, for every $x \in B \cap C$, it is necessarily the case that $x \in A$.

Worksheet 10h and Proposed Solutions

In the course of a game of croquet initiated by the Queen, in which Alice and the cards participated, the King noticed that Alice was talking to someone. The King asked Alice whom she was speaking to, and she introduced him to the Cheshire cat. The King did not like the look of it, but allowed it to kiss his hand. The cat refused the offer and the King found that to be an act of impertinence. The King called the Queen and asked her to get rid of the cat. The Queen, who had only one way of settling all difficulties, called for the cat's beheading, and the King went to fetch the executioner.

After the executioner arrived, a dispute broke out between him and the King regarding the ability to behead the cat, and they presented their arguments to Alice as follows (page 109):

The executioner's argument was, that you couldn't cut off a head unless there was a body to cut it off from: that he had never had to do such a thing before, and he wasn't going to begin at *his* time of life.

The King's argument was, that anything that had a head could be beheaded, and that you weren't to talk nonsense.

1. How can the arguments of the executioner and the King be expressed as conditional statements?

    The executioner's argument: If the creature has no body, then you can't cut off its head.

    The King's argument: If the creature has a head, then you can cut it off.

2. How could Alice settle the question?

We define the following:

a — The creature has a head.
b — The creature's head may be cut off.
c — The creature has a body.

The executioner's assertion: ~c → ~b; that is, b → c.

The King's assertion: a → b.

Thus, Alice could notify the Queen that there is no question, and no dependency exists between the arguments. Moreover, a new assertion may be inferred from the assertions which is consistent with both —a → c; that is: If a creature has a head, then the creature has a body.

Worksheet 10i and Proposed Solutions

Alice got into numerous conversations with the Duchess of Wonderland. Below are excerpts from two such conversations, which took place on different occasions (pages 70 and 113, respectively):

"If everybody minded their own business," the Duchess said in a hoarse growl, "the world would go round a deal faster than it does."

"'Tis so," said the Duchess: "and the moral of that is — 'Oh, 'tis love, 'tis love, that makes the world go round!'"
"Somebody said," Alice whispered, "that it's done by everybody minding their own business!"

It seems that Alice thinks that there is a contradiction between the Duchess' first statement and her second.

a. In order to allow for comparison between the Duchess' statements in each of the two conversations, express them as a condition for the world to go round.

On page 70: A condition for the world to go round is that everyone minds their own business.

On page 113: A condition for the world to go round is that there be love in the world.

b. Do these two statements actually contradict one another?

No, they do not contradict one another. The two conditions may exist simultaneously. Alice's concern was unwarranted.

Worksheet 10j and Proposed Solutions

In Worksheet 3f we were told that the Queen of Hearts made some tarts, and someone ate them. The Queen suspected the Knave of Hearts and put him on trial in the presence of witnesses. Below are excerpts from the trial.

1. The first witness was the Hatter (page 143):

> "Take off your hat," the King said to the Hatter.
> "It isn't mine," said the Hatter.
> "*Stolen!*" the King exclaimed, turning to the jury, who instantly made a memorandum of the fact.
> "I keep them to sell," the Hatter added as an explanation: "I've none of my own. I'm a hatter."

In Wonderland the following statement is TRUE:
(i) The hat on the Hatter's head is not his, and the Hatter has no hats of his own.

In Wonderland the following statement is FALSE:
(ii) The King thinks the hat on the Hatter's head is not stolen.

Which of the following statements is TRUE (there may be more than one right answer):

A. The King thinks that the hat on the Hatter's head is stolen or the Hatter has hats of his own.

B. The hat on the Hatter's head is his, and the King thinks the hat on the Hatter's head is stolen.

C. The hatter has hats of his own or the hat on the Hatter's head is his.

Let us designate the statements as follows:

p: The hat on the Hatter's head is not his.
q: The hatter has no hats of his own.
r: The King thinks the hat on the Hatter's head is not stolen.

Let us construct a truth table for the given information.

	p	q	r	p ∧ q
1	T	T	T	T
2	T	T	F	T
3	T	F	T	F
4	T	F	F	F
5	F	T	T	F
6	F	T	F	F
7	F	F	T	F
8	F	F	F	F

According to the information given in (i), the compound statement p∧q is TRUE (rows 1 and 2), and according to the information given in (ii), the statement r is FALSE (rows 2, 4, 6 and 8).

The row in the table corresponding to both bits of data is row 2, thus we only address that row in considering the three statements whose truth values are in question.

A. Question: Is the statement ~r∨~q TRUE?

We add the relevant details to row 2 above.

p	q	~q	r	~r	~r ∨ ~q
T	T	F	F	T	T

We can now conclude that the statement ~r∨~q is TRUE.

The question may also be answered without using a truth table: The statement ~r is TRUE, since it is given that r is FALSE (as per (ii)). It follows that the compound statement ~r∨~q, constructed using the OR connective, is TRUE, since one of its two component simple statements is TRUE; this suffices to prove the truth of a statement of this kind.

B. *Question: Is the statement ~p∧~r TRUE?*

We add the relevant details to row 2 above.

p	*~p*	*q*	*r*	*~r*	*~p ∧ ~r*
T	F	T	F	T	F

From this we obtain that statement ~p∧~r is FALSE.

The question may also be answered without using a truth table: According to (i), *the compound statement p∧q is TRUE; therefore, its two component simple statements p and q are TRUE as well. Therefore, we obtain that the statement ~p is FALSE. Regardless of the value of ~r, the compound statement ~p∧~r is FALSE, since one of its two simple component statements is FALSE; this suffices to prove the falsity of a statement of this kind.*

C. *Question: Is the statement ~q∨~p TRUE?*

We add the relevant details to row 2 above.

p	*~p*	*q*	*~q*	*~q ∨ ~p*
T	F	T	F	F

We can now conclude that the statement is FALSE.

The question may also be answered without using a truth table. The compound statement p∧q is TRUE; therefore, its two component simple statements p and q are TRUE. It follows that the statements ~p and ~q are FALSE, since the two simple component statements of ~q∨~p are FALSE, and that is the condition by which an OR compound statement is FALSE.

2.   The next witness in the trial was the Duchess's cook, and Alice was the third witness. Alice at that point had grown quite large, and claimed that she knew nothing about the business of stolen tarts (pages 152–153):

> At this moment the King, who had been for some time busily writing in his note-book, called out "Silence!" and read out from his book, "Rule Forty-two. All persons more than a mile high to leave the court."
>
>
>
> Everybody looked at Alice.
> "*I'm* not a mile high," said Alice.
> "You are," said the King.
> "Nearly two miles high," added the Queen.
> "Well, I sha'n't go, at any rate," said Alice: "besides, that's not a regular rule: you invented it just now."
> "It's the oldest rule in the book," said the King.
> "Then it ought to be Number One," said Alice.

Below are three assertions:

(i)   Only kings invent rules.

(ii) All persons more than a mile high leave the court.

(iii) All kings don't leave the court.

Which logical conclusion can be drawn from the three statements together?

Let us designate the statements as follows:

p: The man is the king.
q: The man invents rules.
r: The man is more than a mile high.
s: The man leaves the court.

We express each of the statements as a conditional statement.

(i)   *If a man invents rules, then the man is the king. And in symbolic notation: q → p. Or, equivalently: ~p → ~q.*

(ii)   *If a man is more than a mile high, then he leaves the court. And in symbolic notation: r → s. Or, equivalently: ~s → ~r.*

(iii) *If a man is the king, then he doesn't leave the court. And in symbolic notation: p → ~s. Or, equivalently: s → ~p.*

In order to draw a conclusion from all three statements, let us try to construct a chain of three conditional statements from them (in their current form) that would allow us to use transitivity of the connective IF ... THEN ...

Statements q and r appear only once; therefore they should appear at the ends of the logical chain. We obtain the logical chain: q → p; p → ~s; ~s → ~r. It follows that: q → ~r; stated in words: If a man invents rules then he is not more than a mile high.

3.   When Alice finished speaking, the King immediately requested the jury to consider their verdict. The White Rabbit, who was the herald of the trial, told the King that a paper has just been picked up, a letter written by the prisoner, the Knave of Hearts, and should be used as evidence. Upon unfolding the paper, it turned out that it contained a set of verses. Here is the excerpt (page 154):

"Are they in the prisoner's handwriting?" asked another of the jurymen.
"No, they're not," said the White Rabbit, "and that's the queerest thing about it." (The jury all looked puzzled.)
"He must have imitated somebody else's hand," said the King. (The jury all brightened up again.)

"Please your Majesty," said the Knave, "I didn't write it, and they can't prove that I did: there's no name signed at the end."
"If you didn't sign it," said the King, "that only makes the matter worse. You *must* have meant some mischief, or else you'd have signed your name like an honest man." There was a general clapping of hands at this: it was the first really clever thing the King had said that day.
"That *proves* his guilt, of course," said the Queen: "so, off with——"
"It doesn't prove anything of the sort!" said Alice. "Why, you don't even know what they're about!"

Below are four assertions:

(i)   If a paper is from the Knave of Hearts, then the paper is from the prisoner.

(ii) Every signed paper makes matters worse.

(iii) Every paper that isn't from the Knave of Hearts is not an imitation of somebody else's handwriting.

(iv) Only a paper from someone who is not a prisoner is not signed.

What logical conclusion can be drawn from these four assertions?

Let us designate the statements as follows:

> p: *The paper is from the Knave of Hearts.*
> q: *The paper is from the prisoner.*
> r: *The paper is signed.*
> s: *The paper makes matters worse.*
> t: *The paper is an imitation of somebody else's handwriting.*

We express each of the four statements as a conditional statement.
Assertion (i) *is already expressed as a conditional statement:*
$p \rightarrow q$, or equivalently: $\sim q \rightarrow \sim p$.

Assertion (ii) *may be expressed as follows:*
If a paper is signed, then the paper makes matters worse. Or:
$r \rightarrow s$, or equivalently: $\sim s \rightarrow \sim r$.

Assertion (iii) *may be expressed as follows:*
If a paper is not from the Knave of Hearts, then it is not an imitation of somebody else's handwriting. Or:
$\sim p \rightarrow \sim t$, or, equivalently: $t \rightarrow p$.

Assertion (iv) *may be expressed as follows:*
If a paper is not signed, then it is not from a prisoner. Or:
$\sim r \rightarrow \sim q$, or equivalently: $q \rightarrow r$.

Now let us try to construct a chain of conditional statements, so that from it we can draw conclusions by transitivity.

Statements s and t appear only once; therefore they should appear at the ends of the logical chain. We obtain the logical chain: $t \rightarrow p \rightarrow q \rightarrow r \rightarrow s$. *That is:* $t \rightarrow s$, *or, stated in words: If a paper is an imitation of somebody else's handwriting, then the paper makes matters worse.*

4. The White Rabbit read the verses written on the paper. Here is the excerpt (pages 156–157):

> "That's the most important piece of evidence we've heard yet," said the King, rubbing his hands; "so now let the jury——"
>
> "If any of them can explain it," said Alice, (she had grown so large in the last few minutes that she wasn't a bit afraid of interrupting him,) "I'll give him sixpence. *I* don't believe there's an atom of meaning in it."
> The jury all wrote down on their slates, "*She* doesn't believe there's an atom of meaning in it," but none of them attempted to explain the paper.
> "If there's no meaning in it," said the King, "that saves a world of trouble, you know, as we needn't try to find any. And yet I don't *know*," he went on, spreading out the verses on his knee, and looking at them with one eye; "I seem to see some meaning in them after all.

Below are five assertions:

1. There does not exist evidence acceptable in court that is not important evidence.

2. Evidence that is not written on slates does not save a world of trouble.

3. Evidence taken from verses is acceptable in court.

4. There does not exist important evidence that is meaningless.

5. There does not exist evidence written on slates, unless it is evidence taken from verses.

Which logical conclusion can be drawn from these five assertions?

Let us designate the statements as follows:

p: The evidence is important.
q: The evidence is acceptable in court.
r: The evidence is not written on slates.
s: The evidence saved a world of trouble.
t: The evidence is taken from verses.
u: The evidence is meaningless.

We express each of the five statements as a conditional statement.
Assertion (i) *can be expressed as follows:*
   All evidence acceptable in court is important evidence.
   Or: If evidence is acceptable in court, then it is important
        evidence.
   Or: $q \rightarrow p$, or, equivalently: $\sim p \rightarrow \sim q$.

Assertion (ii) *may be expressed as follows:*
   If evidence is not written on slates, then the evidence does not
   save a world of trouble.
   Or: $r \rightarrow \sim s$, or equivalently: $s \rightarrow \sim r$.

Assertion (iii) *may be expressed as follows:*
   If evidence is taken from verses, then the evidence is acceptable
   in court.
   Or: $t \rightarrow q$, or equivalently: $\sim q \rightarrow \sim t$.

Assertion (iv) *may be expressed as follows:*
   All importance evidence is meaningful evidence.
   Or: If evidence is important, then the evidence is meaningful.
   Or: $p \rightarrow \sim u$, or equivalently: $u \rightarrow \sim p$.

Assertion (v) *may be expressed as follows:*
   All evidence written on slates is evidence taken from verses.
   Or: If evidence is written on slates, then the evidence is taken
        from verses.
   Or: $\sim r \rightarrow t$, or equivalently: $\sim t \rightarrow r$.

Now let us try to construct a chain of conditional statements, so
that from it we can draw conclusions by transitivity.

Statements s and u appear only once; therefore they should appear at the ends of the logical chain. We obtain the logical chain:

$$u \to \sim p \to \sim q \to \sim t \to r \to \sim s$$

That is: $u \to \sim s$, or, stated in words: If evidence is meaningless, then the evidence doesn't save a world of trouble.